Succeed

Eureka Math®
Grade 1
Modules 4–6

Published by Great Minds®.

Copyright © 2018 Great Minds®.

Printed in the U.S.A.
This book may be purchased from the publisher at eureka-math.org
3 4 5 6 7 8 9 10 BAB 25 24 23 22

ISBN 978-1-64054-082-8

G1-M4-M6-S-06.2018

Learn • Practice • Succeed

Eureka Math® student materials for *A Story of Units*® (K–5) are available in the *Learn, Practice, Succeed* trio. This series supports differentiation and remediation while keeping student materials organized and accessible. Educators will find that the *Learn, Practice,* and *Succeed* series also offers coherent—and therefore, more effective—resources for Response to Intervention (RTI), extra practice, and summer learning.

Learn

Eureka Math Learn serves as a student's in-class companion where they show their thinking, share what they know, and watch their knowledge build every day. *Learn* assembles the daily classwork—Application Problems, Exit Tickets, Problem Sets, templates—in an easily stored and navigated volume.

Practice

Each *Eureka Math* lesson begins with a series of energetic, joyous fluency activities, including those found in *Eureka Math Practice.* Students who are fluent in their math facts can master more material more deeply. With *Practice,* students build competence in newly acquired skills and reinforce previous learning in preparation for the next lesson.

Together, *Learn* and *Practice* provide all the print materials students will use for their core math instruction.

Succeed

Eureka Math Succeed enables students to work individually toward mastery. These additional problem sets align lesson by lesson with classroom instruction, making them ideal for use as homework or extra practice. Each problem set is accompanied by a Homework Helper, a set of worked examples that illustrate how to solve similar problems.

Teachers and tutors can use *Succeed* books from prior grade levels as curriculum-consistent tools for filling gaps in foundational knowledge. Students will thrive and progress more quickly as familiar models facilitate connections to their current grade-level content.

Students, families, and educators:

Thank you for being part of the *Eureka Math*® community, where we celebrate the joy, wonder, and thrill of mathematics.

Nothing beats the satisfaction of success—the more competent students become, the greater their motivation and engagement. The *Eureka Math Succeed* book provides the guidance and extra practice students need to shore up foundational knowledge and build mastery with new material.

What is in the Succeed *book?*

Eureka Math Succeed books deliver supported practice sets that parallel the lessons of *A Story of Units*®. Each *Succeed* lesson begins with a set of worked examples, called *Homework Helpers*, that illustrate the modeling and reasoning the curriculum uses to build understanding. Next, students receive scaffolded practice through a series of problems carefully sequenced to begin from a place of confidence and add incremental complexity.

How should Succeed *be used?*

The collection of *Succeed* books can be used as differentiated instruction, practice, homework, or intervention. When coupled with *Affirm*®, *Eureka Math*'s digital assessment system, *Succeed* lessons enable educators to give targeted practice and to assess student progress. *Succeed*'s perfect alignment with the mathematical models and language used across *A Story of Units* ensures that students feel the connections and relevance to their daily instruction, whether they are working on foundational skills or getting extra practice on the current topic.

Where can I learn more about Eureka Math *resources?*

The Great Minds® team is committed to supporting students, families, and educators with an ever-growing library of resources, available at eureka-math.org. The website also offers inspiring stories of success in the *Eureka Math* community. Share your insights and accomplishments with fellow users by becoming a *Eureka Math* Champion.

Best wishes for a year filled with Eureka moments!

Jill Diniz

Jill Diniz
Director of Mathematics
Great Minds

Contents

Module 4: Place Value, Comparison, Addition and Subtraction to 40

Module 5: Identifying, Composing, and Partitioning Shapes

Module 6: Place Value, Comparison, Addition and Subtraction to 100

Grade 1
Module 4

1. Circle groups of 10. Write the number to show the total amount of objects.

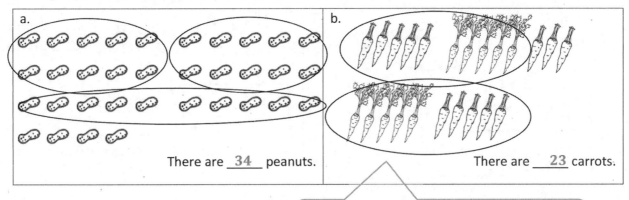

a.

There are __34__ peanuts.

b.

There are __23__ carrots.

> I circle groups of ten. I count the tens first and then the ones. 2 tens 3 ones is 23.

2. Make a number bond to show tens and ones. Circle tens to help. Write the number to show the total amount of objects.

a.

28

20   8

> I think 10, 20, and 8 is 28.

b.

39

30   9

> When I count with ten-sticks, it's much quicker to count. 10, 20, 30, 31, 32, 33, …, 39.

Make or complete a math drawing to show tens and ones. Complete the number bonds.

3.

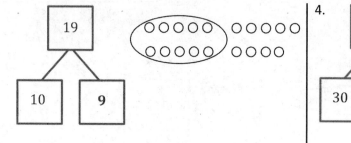

4.

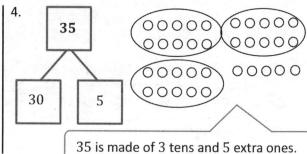

35 is made of 3 tens and 5 extra ones.

Lesson 1:   Compare the efficiency of counting by ones and counting by tens.

Name _____   Date _____

Circle groups of 10. Write the number to show the total amount of objects.

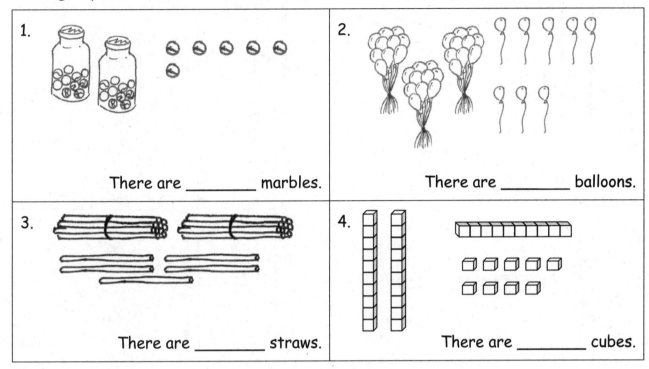

1.

There are _____ marbles.

2.

There are _____ balloons.

3.

There are _____ straws.

4.

There are _____ cubes.

Make a number bond to show tens and ones. Circle tens to help. Write the number to show the total amount of objects.

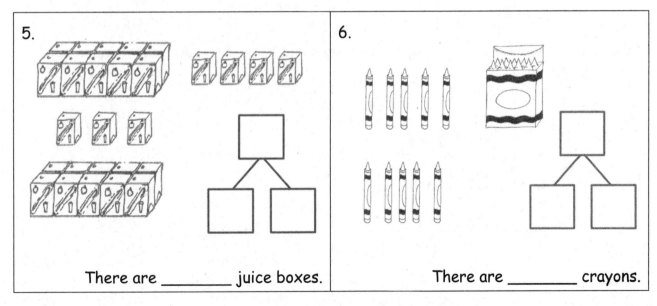

5.

There are _____ juice boxes.

6.

There are _____ crayons.

Lesson 1:   Compare the efficiency of counting by ones and counting by tens.

5

© 2018 Great Minds®. eureka-math.org

Make a number bond to show tens and ones. Circle tens to help. Write the number to show the total amount of objects.

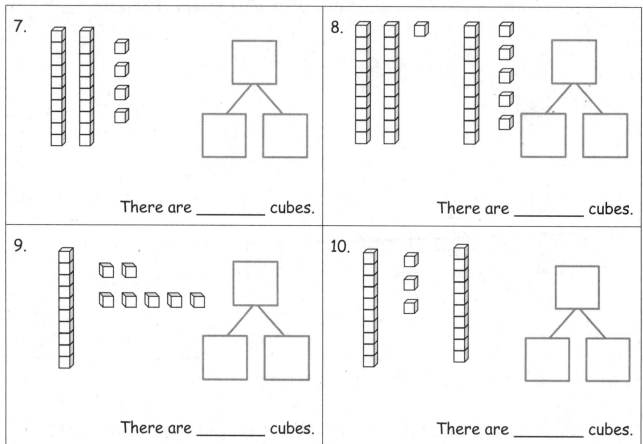

7.

There are _____ cubes.

8.

There are _____ cubes.

9.

There are _____ cubes.

10.

There are _____ cubes.

Make or complete a math drawing to show tens and ones. Complete the number bonds.

11.

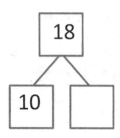

18

10

12.

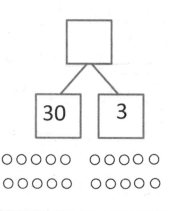

30   3

Lesson 1:   Compare the efficiency of counting by ones and counting by tens.

EUREKA
MATH

Write the tens and ones. Complete the statement.

1.

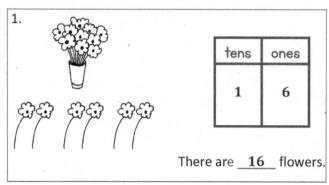

tens	ones
1	6

There are __16__ flowers.

In the number 16, the 1 stands for 1 ten. The 6 stands for 6 ones.

Write the tens and ones. Complete the statement.

2.

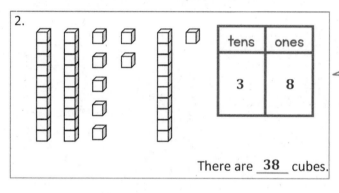

tens	ones
3	8

38 can be separated into 2 parts: 30 and 8. I have 3 ten sticks and 8 extra ones.

There are __38__ cubes.

Write the missing numbers. Say them the regular way and the Say Ten way.

3.

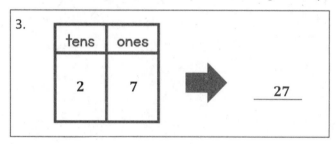

tens	ones
2	7

→ __27__

I look at the place value chart. 2 tens and 7 ones is 27. I can say it the Say Ten way: 2 tens 7.

Lesson 2:   Use the place value chart to record and name tens and ones within a two-digit number.

© 2018 Great Minds®. eureka-math.org

7

4. Choose a number less than 40. Make a math drawing to represent it. Fill in the number bond and place value chart.

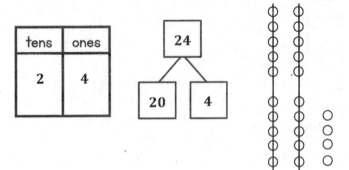

I can make a 5-group column drawing. I draw 2 tens and 4 ones. 24 is 20 and 4.

Lesson 2:   Use the place value chart to record and name tens and ones within a two-digit number.   **EUREKA MATH**

Name _____ Date _____

Write the tens and ones and complete the statement.

1.

tens	ones

There are _____ straws.

2.

tens	ones

There are _____ peanuts.

3.

tens	ones

There are _____ strawberries.

4.

tens	ones

There are _____ beads.

5.

tens	ones

There are _____ apples.

6.

tens	ones

There are _____ carrots.

EUREKA
MATH

Lesson 2: Use the place value chart to record and name tens and ones within a
two-digit number.

© 2018 Great Minds®. eureka-math.org

9

Write the tens and ones. Complete the statement.

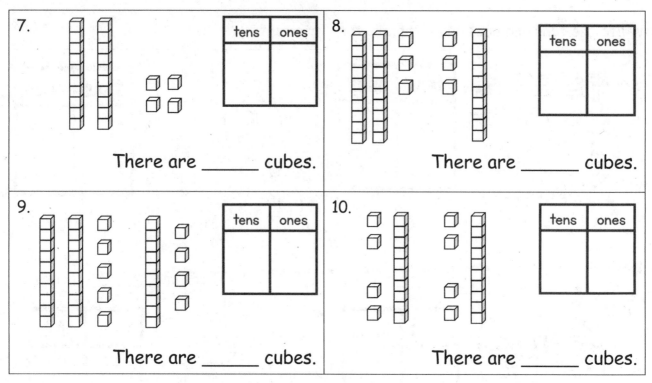

7. There are _____ cubes.

8. There are _____ cubes.

9. There are _____ cubes.

10. There are _____ cubes.

Write the missing numbers. Say them the regular way and the Say Ten way.

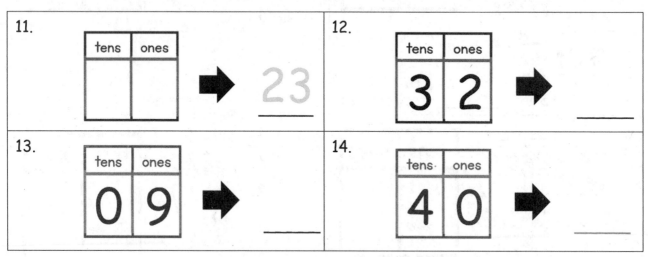

11. 23

12.

13.

14.

15. Choose a number less than 40. Make a math drawing to represent it, and fill in the number bond and place value chart.

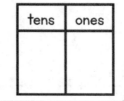

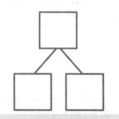

Lesson 2:   Use the place value chart to record and name tens and ones within a two-digit number.

EUREKA MATH®

1. Count as many tens as you can. Complete the statement. Say the numbers and the sentences.

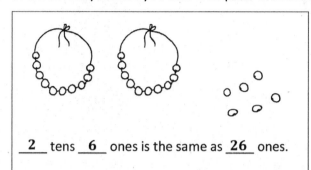

___2___ tens ___6___ ones is the same as ___26___ ones.

I see 26 as 2 tens and 6 extra ones. I count by tens first. 10, 20, and 6 ones is 26.

Fill in the missing numbers.

The number 27 doesn't have 7 ones. It has 27 ones!

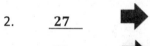

tens	ones
2	7

2. ___27___ ➡ [tens 2 | ones 7] ➡ ___27___ ones

3. ___38___ ➡ 8 ones 3 tens ➡ ___38___ ones

4. ___30___ ➡ ___0___ ones ___3___ tens ➡ 30 ones

There are 38 ones. Or I can say 38 has 3 tens 8 ones. Each ten is made of 10 ones. So, I can count on by tens to get to 30 and then by ones to get to 38.

5. Choose at least one number less than 40. Draw the number in 3 ways:

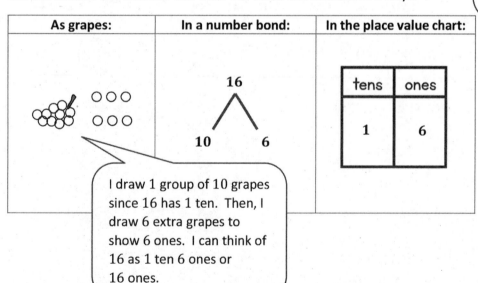

As grapes:	In a number bond:	In the place value chart:
	16 / 10 6	tens 1 | ones 6

I draw 1 group of 10 grapes since 16 has 1 ten. Then, I draw 6 extra grapes to show 6 ones. I can think of 16 as 1 ten 6 ones or 16 ones.

EUREKA MATH

Lesson 3:   Interpret two-digit numbers as either tens and some ones or as all ones.

© 2018 Great Minds®. eureka-math.org

11

Name _____   Date _____

Count as many tens as you can. Complete each statement. Say the numbers and the sentences.

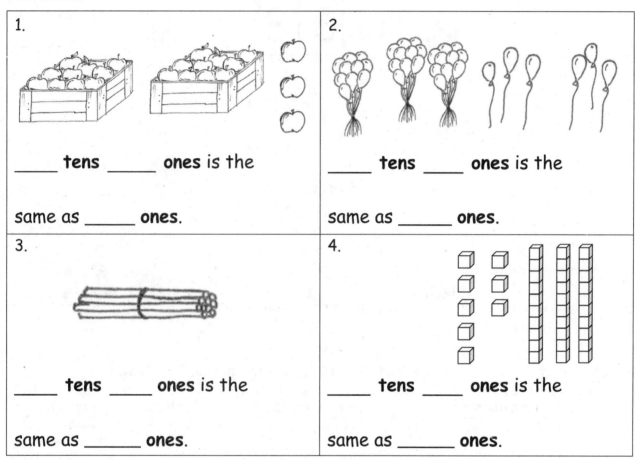

1. ____ **tens** ____ **ones** is the same as _____ **ones**.

2. ____ **tens** ____ **ones** is the same as _____ **ones**.

3. ____ **tens** ____ **ones** is the same as _____ **ones**.

4. ____ **tens** ____ **ones** is the same as _____ **ones**.

Fill in the missing numbers.

5. ____   ➡

tens	ones
2	9

➡   _____ **ones**

EUREKA MATH

Lesson 3: Interpret two-digit numbers as either tens and some ones or as all ones.

© 2018 Great Minds®. eureka-math.org

13

6. **34** ➡ ____ tens ____ ones ➡ ____ ones

7. ____ ➡ | tens | ones |
           | 3 | 8 | ➡ ____ ones

8. ____ ➡ 9 ones 3 tens ➡ ____ ones

9. ____ ➡ ____ ones ____ tens ➡ **40** ones

10. Choose at least one number less than 40. Draw the number in 3 ways:

As grapes:	In a number bond:	In the place value chart:
	∧	tens | ones

Lesson 3:   Interpret two-digit numbers as either tens and some ones or as all ones.

© 2018 Great Minds®. eureka-math.org

EUREKA MATH

1.   Fill in the number bond, or write the tens and ones. Complete the addition sentences.

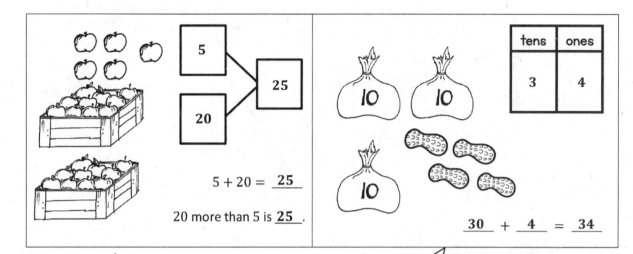

$5 + 20 = $ __25__

20 more than 5 is __25__ .

$30 + 4 = 34$

I can make a number bond that shows the tens and ones. I can break apart 25 into 20 and 5.

3 tens 4 ones is the same as the number 34. 3 is the digit in the tens place, and 4 is the digit in the ones place.

Lesson 4:   Write and interpret two-digit numbers as addition sentences that combine tens and ones.

15

EUREKA
MATH®

2. Match the pictures with the words.

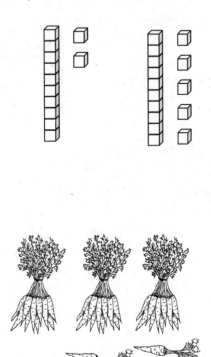

This statement combines tens and ones!

2 more than 30 is __32__.

$20 + 7 = 27$

I can write a number sentence with the tens first, or I can write it with the ones first, like $7 + 20 = 27$. One number tells how many tens there are, and the other tells how many ones there are.

Lesson 4:   Write and interpret two-digit numbers as addition sentences that combine tens and ones.

EUREKA MATH

Name _____ Date _____

Fill in the number bond, or write the tens and ones. Complete the addition sentences.

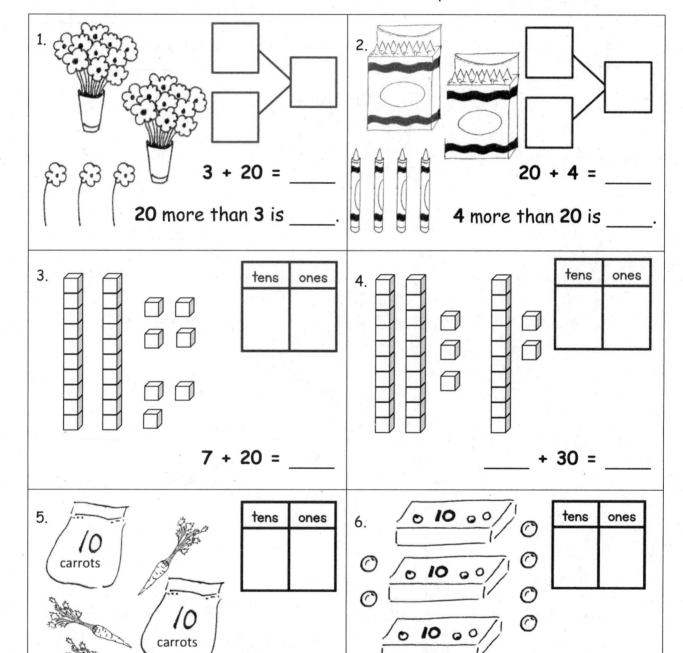

1.

3 + 20 = _____

20 more than 3 is _____.

2.

20 + 4 = _____

4 more than 20 is _____.

3.

tens	ones

7 + 20 = _____

4.

tens	ones

_____ + 30 = _____

5.

tens	ones

10 carrots

10 carrots

20 + _____ = _____

6.

tens	ones

10

10

10

_____ + _____ = _____

EUREKA MATH

Lesson 4: Write and interpret two-digit numbers as addition sentences that combine tens and ones.

17

© 2018 Great Minds®. eureka-math.org

Match the pictures with the words.

7.

•   • | 1 and 30 make _____. |

8.

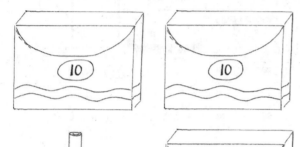

•   • | 8 + 30 = _____. |

9.

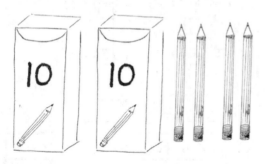

•   • | 2 more than 10 is _____. |

10.

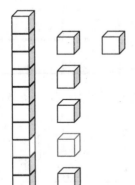

•   • | 20 + 4 = _____. |

Lesson 4:   Write and interpret two-digit numbers as addition sentences that combine tens and ones.

EUREKA
MATH

Draw quick tens and ones to show the number. Then draw 1 more or 10 more.

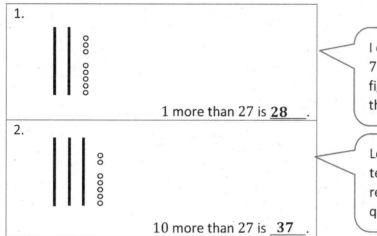

1.

1 more than 27 is **28**.

I can show 27 with 2 quick tens and 7 ones in a 5-group column. To figure out 1 more, I add 1 circle to the ones, so 7 ones becomes 8 ones.

2.

10 more than 27 is **37**.

Look at how quickly I can draw 37. A quick ten is a line that holds 10 beads! It represents a ten. I can draw one more quick ten to show 10 more than 27.

Draw quick tens and ones to show the number. Cross off (x) to show 1 less or 10 less.

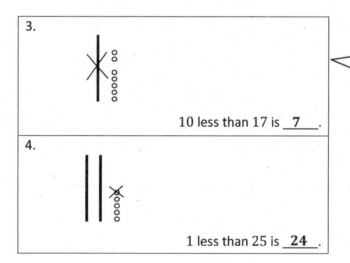

3.

10 less than 17 is **7**.

I can cross out a quick ten when I want to show 10 less than 17. Now, there are no tens and 7 ones.

4.

1 less than 25 is **24**.

Match the words to the picture that shows the right amount.

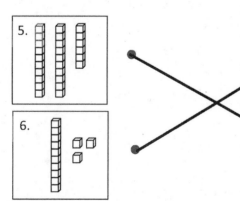

5.

6.

10 less than 23

10 more than 16

The digit in the tens place changes when I think of 10 more than 16. The new number is 26. That's 2 tens 6 ones.

Name _____   Date _____

Draw quick tens and ones to show the number. Then, draw 1 more or 10 more.

1. 1 more than 38 is _____.	2. 10 more than 38 is _____.
3. 1 more than 35 is _____.	4. 10 more than 35 is _____.

Draw quick tens and ones to show the number. Cross off (x) to show 1 less or 10 less.

5. 10 less than 23 is _____.	6. 1 less than 23 is _____.
7. 10 less than 31 is _____.	8. 1 less than 31 is _____.

Lesson 5:   Identify 10 more, 10 less, 1 more, and 1 less than a two-digit number.   **21**

© 2018 Great Minds®. eureka-math.org

Match the words to the picture that shows the right amount.

9.

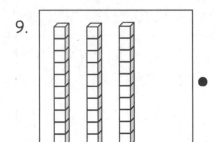

•

• 1 less than 30.

10.

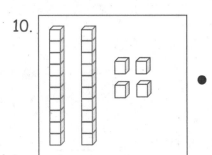

•

• 1 more than 23.

11.

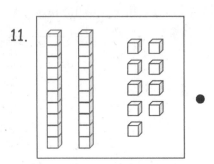

•

• 10 less than 36.

12.

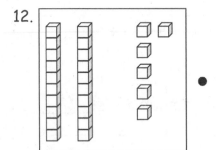

•

• 10 more than 20.

Lesson 5:   Identify 10 more, 10 less, 1 more, and 1 less than a two-digit number.

© 2018 Great Minds®. eureka-math.org

EUREKA MATH

Fill in the place value chart and the blanks.

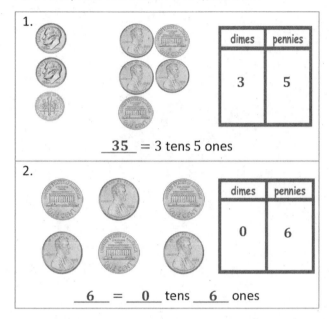

1 dime has the same value as 10 pennies, but it's just 1 coin. 3 dimes and 5 pennies equal 3 tens 5 ones. That's 35 cents!

I don't see any tens because there are no dimes. The value of 6 pennies is 6 cents.

Fill in the blank. Draw or cross off tens or ones as needed.

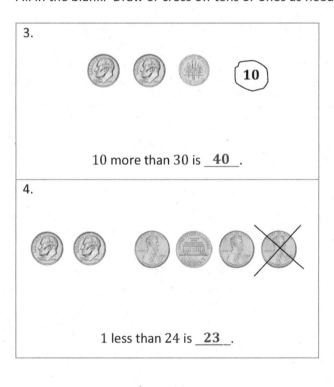

I can draw 1 more dime since I want to show 10 more. So, 3 tens changes to 4 tens. 30 cents + 10 cents = 40 cents.

When I cross off 1 penny, I have 1 less, or 23 cents. I could write this in my place value chart as 2 tens 3 ones.

Lesson 6:   Use dimes and pennies as representations of tens and ones.

23

Name _____   Date _____

Fill in the place value chart and the blanks.

1.
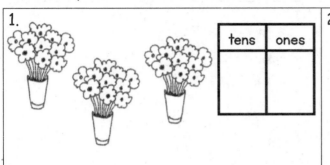

tens	ones

30 = _____ tens

2.
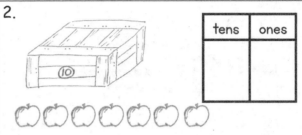

tens	ones

17 = _____ ten and _____ ones

3.

dimes	pennies

_____ = 2 tens 2 ones

4.
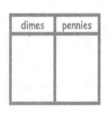

dimes	pennies

_____ = 3 tens 3 ones

5.

dimes	pennies

_____ = _____ tens _____ ones

6.

dimes	pennies

_____ = _____ tens _____ ones

7.

tens	ones

_____ = _____ ten _____ ones

8.

tens	ones

_____ tens _____ ones = _____

EUREKA MATH

Lesson 6:   Use dimes and pennies as representations of tens and ones.

25

© 2018 Great Minds®. eureka-math.org

Fill in the blank. Draw or cross off tens or ones as needed.

10 more than 25 is __35__

9. 1 more than 12 is _____.	10. 10 more than 3 is _____.
11. 10 more than 22 is _____.	12. 1 more than 22 is _____.
13. 1 less than 39 is _____.	14. 10 less than 39 is _____.
15. 10 less than 33 is _____.	16. 1 less than 33 is _____.

Lesson 6:   Use dimes and pennies as representations of tens and ones.

EUREKA
MATH®

Write the number, and circle the set that is *greater* in each pair. Say a statement to compare the two sets.

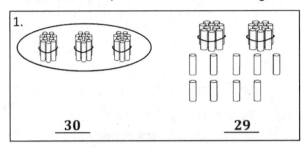

1.

____30____   ____29____

> I look at the tens place first to find the number that is greater. 3 tens is more than 2 tens. So, 30 is greater than 29.

Circle the number that is *greater* for each pair.

2.

3 tens 9 ones   (4 tens 8 ones)

> 4 tens is greater than 3 tens, so 48 is greater than 39.

Write the number, and circle the set that is *less* in each pair. Say a statement to compare the two sets.

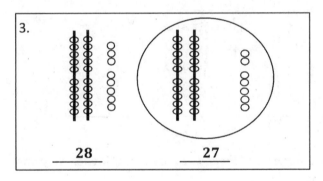

3.

____28____   ____27____

> First, I look at the tens place and both numbers have 2 tens. Next, I look at the ones place, and 7 ones is less than 8 ones. So, 27 is less than 28.

Lesson 7:   Compare two quantities, and identify the greater or lesser of the two given numerals.

4. Write the value, and circle the set of coins that has *less* value.

14 cents

22 cents

> The first set has 5 coins, and the second set has 4 coins, but you have to look at the values! Dimes and pennies are like tens and ones. So, 1 ten 4 ones is less than 2 tens 2 ones.

5. Maddox and Caroline are playing cards. If Caroline's total has 29 ones and Maddox's total is 26, whose total is less? Draw a math drawing to explain how you know.

> Hey, 29 ones is also 2 tens 9 ones! I can draw a picture and just compare ones!

C   *M*

Maddox's total is less. I know because they both have 2 tens, so I looked at the ones. Maddox only has 6 ones, and Caroline has 9 ones. So, Maddox has less.

     Lesson 7:   Compare two quantities, and identify the greater or lesser of the two given numerals.

© 2018 Great Minds®. eureka-math.org

EUREKA MATH®

Name _____ Date _____

Write the number, and circle the set that is *greater* in each pair. Say a statement to compare the two sets.

1.
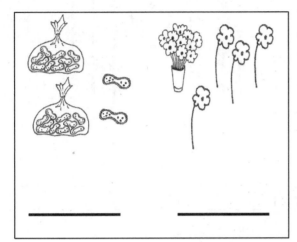

_____   _____

2.
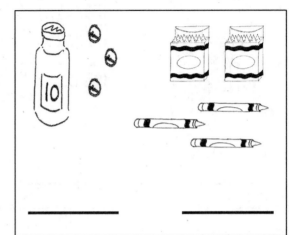

_____   _____

Circle the number that is *greater* for each pair.

3.

3 tens 8 ones	3 tens 9 ones

4.

25	35

5. Write the value and circle the set of coins that has *greater* value.

_____   _____

EUREKA MATH

Lesson 7:   Compare two quantities, and identify the greater or lesser of the two given numerals.

© 2018 Great Minds®. eureka-math.org

29

Write the number, and circle the set that is *less* in each pair. Say a statement to compare the two sets.

6.

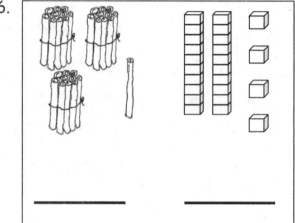

_____   _____

7.

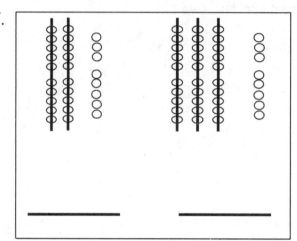

_____   _____

Circle the number that is *less* for each pair.

8.

| 2 tens 7 ones | 3 tens 7 ones |

9.

| 22 | 29 |

10. Write the value and circle the set of coins that has *less* value.

_____   _____

30   Lesson 7:   Compare two quantities, and identify the greater or lesser of the two
                            given numerals.

© 2018 Great Minds®. eureka-math.org

EUREKA
MATH

11. Katelyn and Johnny are playing comparison with cards. They have recorded the totals for each round. For each round, circle the total that won the cards, and write the statement. The first one is done for you.

ROUND 1: The total that is **greater** wins.

Katelyn's Total	Johnny's Total
16	19

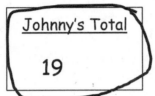

19 is greater than 16.

a. ROUND 2: The total that is **less** wins.

Katelyn's Total	Johnny's Total
27	24

_____

b. ROUND 3: The total that is **greater** wins.

Katelyn's Total	Johnny's Total
32	22

_____

c. ROUND 4: The total that is **less** wins.

Katelyn's Total	Johnny's Total
29	26

_____

d. If Katelyn's total is 39, and Johnny's total has 3 tens 9 ones, who would have a greater total? Draw a math drawing to explain how you know.

EUREKA MATH   Lesson 7:   Compare two quantities, and identify the greater or lesser of the two   31
                                given numerals.

© 2018 Great Minds®. eureka-math.org

Word Bank

is greater than
is less than
is equal to

1. Draw the numbers using quick tens and circles. Use the phrases from the word bank to complete the sentence frames to compare the numbers.

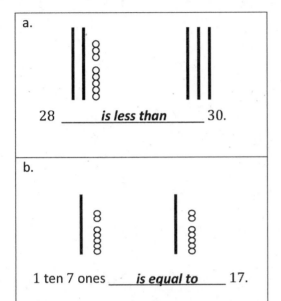

    28 ___*is less than*___ 30.

b.

1 ten 7 ones ___*is equal to*___ 17.

> I look at the digit in the tens place first to compare the numbers! Even though there are 8 ones in 28, that's still less than a ten. I read from left to right: 28 is less than 30.

> 3 tens 3 ones is 33. Both numbers have 3 tens, but 3 ones is less than 4 ones. So, 3 tens 3 ones is less than 34.

2. Circle the numbers that are *less* than 34.

    (29)   3 tens 5 ones   4 tens   (31)   (3 tens 3 ones)

3. Write the numbers in order from *greatest* to *least*.

    24
    12
        40
    16

> I read the numbers from left to right. 40 is greater than 24. 24 is greater than 16....

___40___   ___24___   ___16___   ___12___

Where would the number 38 go in this order? Use words or rewrite the numbers to explain.

    40   **38**   **24**   **16**   **12**

> I put 38 between 40 and 24. 38 is less than 40, and 38 is greater than 24. Look at the tens: 4 tens, 3 tens, 2 tens!

Name _____   Date _____

1. Draw the numbers using quick tens and circles. Use the phrases from the word bank to complete the sentence frames to compare the numbers. The first one has been done for you.

Word Bank

is greater than

is less than

is equal to

a. 20 ‖ 30 ‖‖ 20 ___is less than___ 30	b. 14 22 14 _____ 22
c. 15 1 ten 5 ones 15 _____ 1 ten 5 ones	d. 39 29 39 _____ 29
e. 31 13 31 _____ 13	f. 23 33 23 _____ 33

2. Circle the numbers that are *greater* than 28.

   32   29   2 tens 8 ones   4 tens   18

3. Circle the numbers that are *less* than 31.

   29   3 tens 6 ones   3 tens   13   3 tens 9 ones

4. Write the numbers in order from *least* to *greatest*.

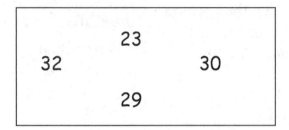

Where would the number 27 go in this order? Use words or rewrite the numbers to explain.

5. Write the numbers in order from *greatest* to *least*.

	40	
13		30
	31	

_____ _____ _____ _____

Where would the number 23 go in this order? Use words or rewrite the numbers to explain.

6. Use the digits 9, 4, 3, and 2 to make 4 different two-digit numbers less than 40. Write them in order from *least* to *greatest*.

9	3	4	2

Examples: 34, 29,...

Lesson 8:   Compare quantities and numerals from left to right.

© 2018 Great Minds®. eureka-math.org

EUREKA MATH

1. Write the numbers in the blanks so that the alligator is eating the greater number. Read the number sentence, using *is greater than, is less than,* or *is equal to.* Remember to start with the number on the left.

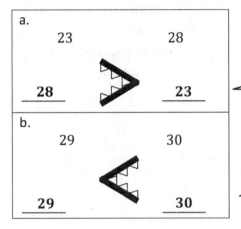

a.

    23   28

    __28__   >   __23__

> I remember to read starting with the number on the left. So, 28 is greater than 23. I know because 2 tens 8 ones is greater than 2 tens 3 ones.

b.

    29   30

    __29__   <   __30__

> 29 is less than 30. 30 is 3 tens! The alligator wants to eat the bigger number!

2. Complete the charts so that the alligator is eating a *greater* number.

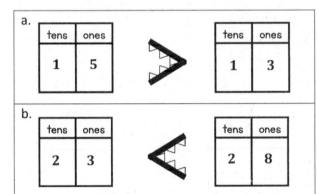

a.

tens	ones		tens	ones
1	5	>	1	3

> I read the number sentence as 15 is greater than 13. Both numbers have 1 ten, but 5 ones is bigger than 3 ones, so the alligator eats the number 15.

b.

tens	ones		tens	ones
2	3	<	2	8

> I write 8 in the ones place, so the alligator eats the number 28. I can read the number sentence as 23 is less than 28. I could also write 4, 5, 6, 7, 8, or 9 ones, too!

3. Compare each set of numbers by matching to the correct alligator or phrase to make a true number sentence. Check your work by reading the sentence from left to right.

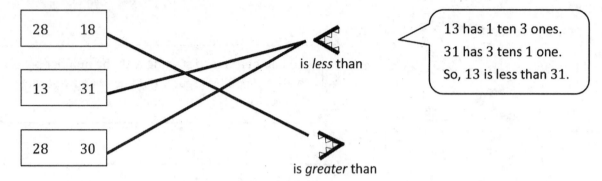

13 has 1 ten 3 ones.
31 has 3 tens 1 one.
So, 13 is less than 31.

EUREKA
MATH

Name _____   Date _____

1. Write the numbers in the blanks so that the alligator is eating the greater number. Read the number sentence, using *is greater than, is less than,* or *is equal to.* Remember to start with the number on the left.

a. 10 20	b. 15 17	c. 24 22
____ > ____	____ < ____	____ > ____
d. 29 30	e. 39 38	f. 39 40
____ > ____	____ < ____	____ < ____

2. Complete the charts so that the alligator is eating a *greater* number.

a.

tens	ones
1	8

>

tens	ones
1	

b.

tens	ones
2	4

<

tens	ones
	3

c.

tens	ones

>

tens	ones

d.

tens	ones
2	3

>

tens	ones
2	

e.

tens	ones

<

tens	ones

f.

tens	ones
1	7

>

tens	ones
	7

Compare each set of numbers by matching to the correct alligator or phrase to make a true number sentence. Check your work by reading the sentence from left to right.

3.

| 16 | 17 |

| 31 | 23 |

| 35 | 25 |

| 12 | 21 |

| 22 | 32 |

| 29 | 30 |

| 39 | 40 |

<

is *less* than

>

is *greater* than

Lesson 9:   Use the symbols >, =, and < to compare quantities and numerals.

EUREKA
MATH

Use the symbols to compare the numbers. Fill in the blank with <, >, or = to make a true number sentence.
Complete the number sentence with a phrase from the word bank.

Word Bank

is greater than

is less than

is equal to

a.

21 (>) 12

21 __is greater than__ 12.

> Both of these numbers have the same digits, but they are in different positions. That means they have a different value. 2 tens 1 one is greater than 1 ten 2 ones!

b.

3 tens (<) 32

3 tens __is less than__ 32.

> I put the less than sign between 3 tens and 32. 3 tens is 30. The smaller end points to the smaller number!

c.

2 tens 8 ones (<) 29

2 tens 8 ones __is less than__ 29.

> There are more ones in 29 than in 2 tens 8 ones, or 28. The symbol is open on the side that the alligator likes to eat! But I still read it from left to right!

d.

19 (=) 1 ten 9 ones

19 __is equal to__ 1 ten 9 ones.

EUREKA MATH

Name _____   Date _____

Use the symbols to compare the numbers. Fill in the blank with <, >, or = to make a true number sentence. Complete the number sentence with a phrase from the word bank.

Word Bank

is greater than
is less than
is equal to

40 (>) 20
40 is greater than 20.

18 (<) 20
18 is less than 20.

a.

17 ( ) 13

17 _____ 13

b.

23 ( ) 33

23 _____ 33

c.

36 ( ) 36

36 _____ 36

d.

25 ( ) 32

25 _____ 32

e.

38 ( ) 28

38 _____ 28

f.

32 ( ) 23

32 _____ 23

g.

1 ten 5 ones   ◯   14

1 ten 5 ones _____ 14

h.

3 tens   ◯   30

3 tens _____ 30

i.

29   ◯   2 tens 7 ones

29 _____ 2 tens 7 ones

j.

19   ◯   2 tens 3 ones

19 _____ 2 tens 3 ones

k.

3 tens 1 one   ◯   13

3 tens 1 one _____ 13

l.

35   ◯   3 tens 5 ones

35 _____ 3 tens 5 ones

m.

2 tens 3 ones   ◯   32

2 tens 3 ones _____ 32

n.

3 tens   ◯   36

3 tens _____ 36

o.

29   ◯   3 tens 9 ones

29 _____ 3 tens 9 ones

p.

4 tens   ◯   39

4 tens _____ 39

Lesson 10:   Use the symbols >, =, and < to compare quantities and numerals.

EUREKA MATH

Draw a number bond, and complete the number sentences to match the pictures.

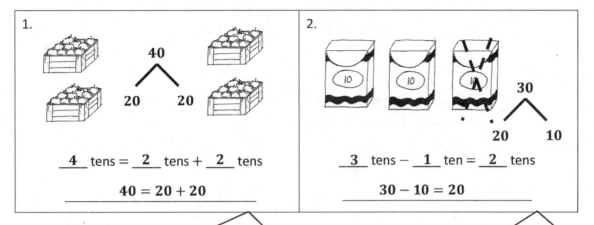

1.

_**4**_ tens = _**2**_ tens + _**2**_ tens

40 = 20 + 20

2.

_**3**_ tens − _**1**_ ten = _**2**_ tens

30 − 10 = 20

> I can say the number sentence with place value units, so 4 tens = 2 tens + 2 tens. That's the unit way. Or I can just write the numbers the regular way, so 40 = 20 + 20.

> The number bond shows 3 tens on top with 2 tens and 1 ten as the parts. The X shows that I take away 1 ten. The subtraction sentences match.

Draw quick tens and a number bond to help you solve the number sentences.

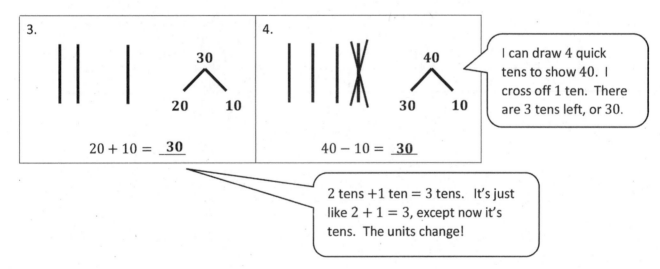

3.

20 + 10 = _**30**_

4.

40 − 10 = _**30**_

> I can draw 4 quick tens to show 40. I cross off 1 ten. There are 3 tens left, or 30.

> 2 tens +1 ten = 3 tens. It's just like 2 + 1 = 3, except now it's tens. The units change!

Add or subtract.

5.   4 tens − 3 tens = __1 *ten*__

6.   __40__ = 10 + 30

I can think of the simpler problem,
$4 = 1 + 3$, to help me solve.

7.   **20** − 20 = __0__

EUREKA
MATH

Name _____ Date _____

Draw a number bond, and complete the number sentences to match the pictures.

1.

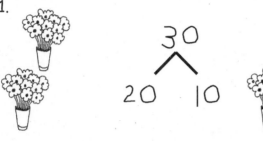

__2__ tens + __1__ ten = __3__ tens

20 + 10 = 30

2.

____ tens = ____ ten + ____ tens

_____

3.

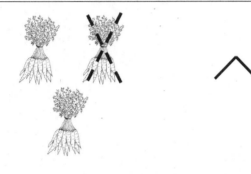

____ tens - ____ ten = ____ tens

_____

4.

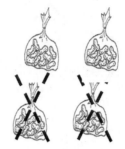

____ tens - ____ tens = ____ tens

_____

5.

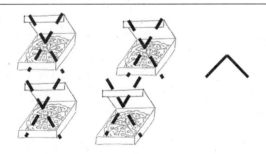

____ tens - ____ tens = ____ tens

_____

6.

____ tens + ____ tens = ____ tens

_____

Draw quick tens and a number bond to help you solve the number sentences.

7. 10 + 20 = _____	8. 30 − 10 = _____
9. 20 − 10 = _____	10. 30 + 10 = _____

Add or subtract.

11. 2 tens + 1 ten = _____   12. 20 + 20 = _____   13. 40 − 10 = _____

14. _____ = 20 + 10   15. 3 tens − 2 tens = _____   16. 20 − 10 = _____

17. 10 − 10 = _____   18. _____ = 30 + 10   19. 40 − 30 = _____

Lesson 11:   Add and subtract tens from a multiple of 10.

EUREKA
MATH

1. Fill in the missing numbers to match the picture. Write the matching number bond.

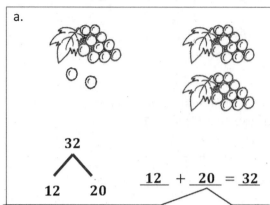

a.

32

12   20

__12__ + __20__ = 32

1 ten 2 ones + 2 tens = 3 tens 2 ones. The digit in the tens place changes because I add 2 tens. The ones stay the same.

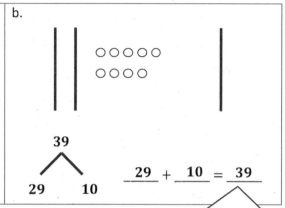

b.

39

29   10

__29__ + __10__ = __39__

1 ten more than 2 tens is 3 tens. That's why there is a 3 in the tens place. There are still 9 ones.

2. Draw using quick tens and ones. Complete the number bond and the number sentence.

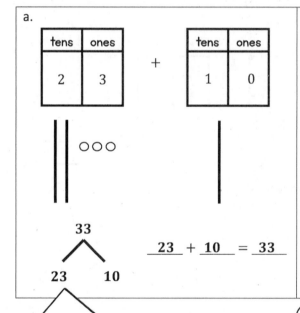

a.

tens	ones
2	3

+

tens	ones
1	0

33

23   10

__23__ + __10__ = __33__

The number bond shows how I change 23 to make 33. I add 1 ten.

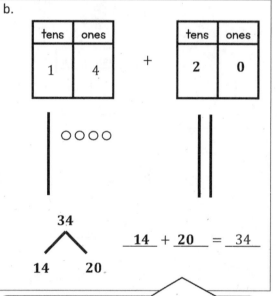

b.

tens	ones
1	4

+

tens	ones
2	0

34

14   20

__14__ + __20__ = __34__

If 34 is the whole and 14 is one part, I can add 2 tens to make 34. 2 tens is the same as 20. 14 plus 20 equals 34.

3. Use arrow notation to solve.

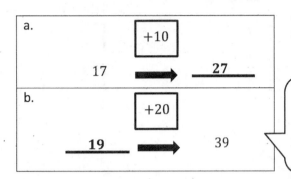

a.

+10

17 ➡ **27**

b.

+20

19 ➡ 39

> I can think: What number plus 2 tens will give me 3 tens 9 ones? 1 ten 9 ones plus 2 tens equals 3 tens 9 ones! So, 19 is the number.

4. Use the dimes and pennies to complete the place value charts.

a.

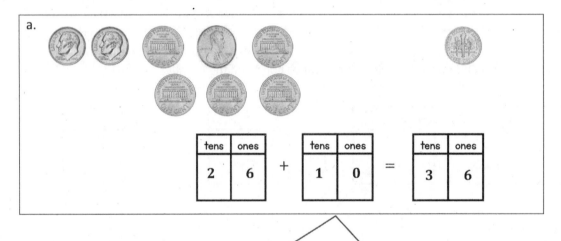

tens	ones		tens	ones		tens	ones
2	6	+	1	0	=	3	6

> 2 dimes and 6 pennies make 2 tens 6 ones. When I add 1 dime, I add 1 ten. Now, there are 3 tens all together. The number sentence is $26 + 10 = 36$.

EUREKA MATH

Name _____ Date _____

Fill in the missing numbers to match the picture. Complete the number bond to match.

1.

$\wedge$

20 + 13 = _____

2.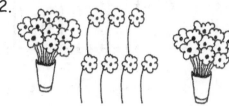

$\wedge$

17 + _____ = _____

3.

$\wedge$

_____ + _____ = _____

4.

$\wedge$

_____ + _____ = _____

Draw using quick tens and ones. Complete the number bond and the number sentence.

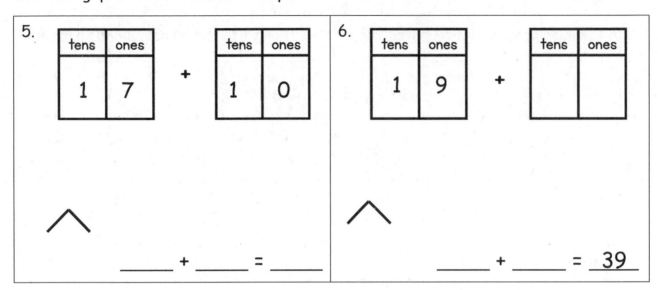

5.

tens	ones
1	7

+

tens	ones
1	0

_____ + _____ = _____

6.

tens	ones
1	9

+

tens	ones

_____ + _____ = __39__

Use arrow notation to solve.

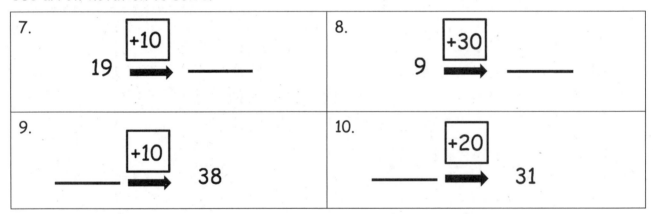

7.
$19 \xrightarrow{+10}$ _____

8.
$9 \xrightarrow{+30}$ _____

9.
_____ $\xrightarrow{+10}$ 38

10.
_____ $\xrightarrow{+20}$ 31

Use the dimes and pennies to complete the place value charts.

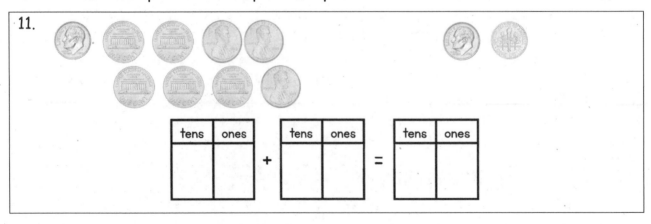

11.

tens	ones

+

tens	ones

=

tens	ones

EUREKA
MATH®

1. Use quick tens and ones to complete the place value chart and number sentence.

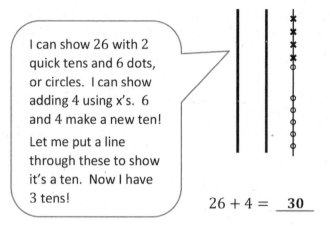

I can show 26 with 2 quick tens and 6 dots, or circles. I can show adding 4 using x's. 6 and 4 make a new ten!

Let me put a line through these to show it's a ten. Now I have 3 tens!

tens	ones
3	0

$26 + 4 = \underline{\ 30\ }$

2. Draw quick tens, ones, and number bonds to solve. Complete the place value chart.

$25 + 5 = \underline{\ 30\ }$

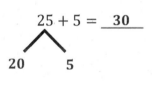

20   5

tens	ones
3	0

25 is made of 20 and 5. I can add 5 and 5 to make 10. Then I know that $20 + 10 = 30$. That's 3 tens.

3. Solve. You may draw quick tens and ones or number bonds to help.

$37 + 3 = \underline{\ 40\ }$

I know this one in my head. 3 more than 37 is 40. I am making the next ten when I add 3 to 37.

EUREKA MATH

Lesson 13:   Use counting on and the make ten strategy when adding across a ten.

53

© 2018 Great Minds®. eureka-math.org

Name _____ Date _____

Use quick tens and ones to complete the place value chart and number sentence.

1.				2.			

1.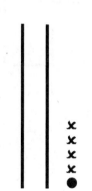

tens	ones

21 + 4 = _____

2.
tens	ones

21 + 8 = _____

3.
tens	ones

25 + 4 = _____

4.
tens	ones

25 + 5 = _____

5.
tens	ones

33 + 3 = _____

6.
tens	ones

33 + 7 = _____

EUREKA MATH®

Lesson 13:   Use counting on and the make ten strategy when adding across a ten.

55

© 2018 Great Minds®. eureka-math.org

Draw quick tens, ones, and number bonds to solve. Complete the place value chart.

7.

26 + 2 = _____

tens	ones

8.

36 + 3 = _____

tens	ones

9.

26 + 4 = _____

tens	ones

10.

24 + 6 = _____

tens	ones

11. Solve. You may draw quick tens and ones or number bonds to help.

a. 22 + 7 = _____   b. 22 + 8 = _____   c. 32 + 8 = _____

Lesson 13:   Use counting on and the make ten strategy when adding across a ten.

EUREKA MATH

1. Use the pictures, or draw quick tens and ones. Complete the number sentence and place value chart.

I can use 2 quick tens and 9 dots, or circles, to show 29. I only need one more to make a new ten. As I add 5, the first x makes a new ten. I start a new column as I draw 4 more x's. I can draw a line through the new ten I made. Now I can see easily that I have 3 tens and 4 ones.

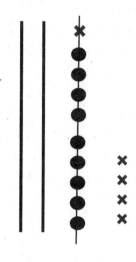

tens	ones
3	4

$29 + 5 =$ __34__

2. Make a number bond to solve. Show your thinking with number sentences or the arrow way. Complete the place value chart.

$18 + 5 =$ __23__

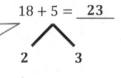

tens	ones
2	3

I need 2 more to get to 20 from 18. I can break apart 5 into 2 and 3. $18 + 2 = 20$. Then $20 + 3 = 23$.

Here are my number sentences to show my thinking.

$$18 + 2 = 20$$
$$20 + 3 = 23$$

$18 \xrightarrow{+2} 20 \xrightarrow{+3} 23$

I can use the arrow way to show my thinking too! I start at 18. I add 2 to get to 20. Then, I add 3 more to get to 23.

EUREKA MATH®

Name _____   Date _____

Use the pictures or draw quick tens and ones. Complete the number sentence and place value chart.

1.	2.	3.
15 + 3 = _____	15 + 5 = _____	15 + 6 = _____

4.	5.	6.
28 + 2 = _____	28 + 4 = _____	28 + 7 = _____

7.	8.	9.
17 + 3 = _____	17 + 7 = _____	27 + 7 = _____

Lesson 14:   Use counting on and the make ten strategy when adding across a ten.

59

EUREKA
MATH

© 2018 Great Minds®. eureka-math.org

Make a number bond to solve. Show your thinking with number sentences or the arrow way. Complete the place value chart.

10.		
13 + 6 = _____	tens	ones

11.		
13 + 7 = _____	tens	ones

12.		
25 + 5 = _____	tens	ones

13.		
25 + 8 = _____	tens	ones

14.		
24 + 8 = _____	tens	ones

15.		
23 + 9 = _____	tens	ones

Lesson 14:   Use counting on and the make ten strategy when adding across a ten.

EUREKA
MATH

1. Solve the problems.

 $9 + 5 =$ __14__

> 9 plus 5 is 14.
> That one's easy.

     $19 + 5 =$ __24__

> 19 plus 5 is
> just 10 more.
> That's 24.

$29 + 5 =$ __34__

> 29 plus 5 is 10
> more again.
> That's 34.

2. Use the first number sentence in each set to help you solve the other problems.

    a. $3 + 8 =$ __11__

    b. $13 + 8 =$ __21__

    c. $23 + 8 =$ __31__

3. Solve the problems. Show the 1-digit addition sentence that helped you solve.

    $18 + 4 =$ __22__   ____ $8 + 4 = 12$ ____

    > I can use $8 + 4$ to help me solve
    > $18 + 4$. I know that $8 + 4 = 12$.
    > $18 + 4$ has 1 more ten. That's 22.

EUREKA MATH

Lesson 15:   Use single-digit sums to support solutions for analogous sums to 40.

61

© 2018 Great Minds®. eureka-math.org

Name _____ Date _____

Solve the problems.

1.

$5 + 4 =$ _____

2.
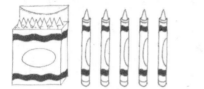

$15 + 4 =$ _____

3.
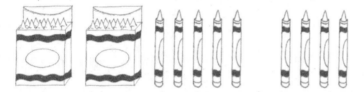

$25 + 4 =$ _____

4.
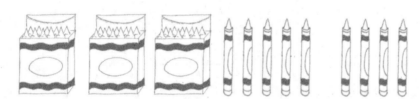

$35 + 4 =$ _____

5.

$8 + 4 =$ _____

6.
  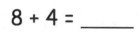

$18 + 4 =$ _____

7.
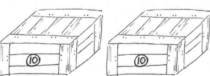 

$28 + 4 =$ _____

Lesson 15: Use single-digit sums to support solutions for analogous sums to 40.

63

EUREKA MATH

Use the first number sentence in each set to help you solve the other problems.

8. a. 5 + 2 = _____ b. 15 + 2 = _____ c. 25 + 2 = _____ d. 35 + 2 = _____	9. a. 5 + 5 = _____ b. 15 + 5 = _____ c. 25 + 5 = _____ d. 35 + 5 = _____
10. a. 2 + 7 = _____ b. 12 + 7 = _____ c. 22 + 7 = _____	11. a. 7 + 4 = _____ b. 17 + 4 = _____ c. 27 + 4 = _____
12. a. 8 + 7 = _____ b. 18 + 7 = _____ c. 28 + 7 = _____	13. a. 3 + 9 = _____ b. 13 + 9 = _____ c. 23 + 9 = _____

Solve the problems. Show the 1-digit addition sentence that helped you solve.

14. 24 + 5 = _____   _____

15. 24 + 7 = _____   _____

EUREKA
MATH

1. Draw quick tens and ones to help you solve the addition problems.

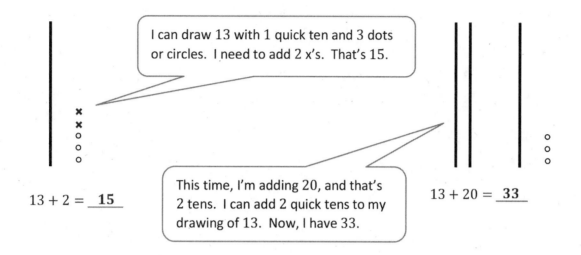

I can draw 13 with 1 quick ten and 3 dots or circles. I need to add 2 x's. That's 15.

This time, I'm adding 20, and that's 2 tens. I can add 2 quick tens to my drawing of 13. Now, I have 33.

$13 + 2 =$ __15__

$13 + 20 =$ __33__

2. Make a number bond, or use the arrow way to solve the addition problems.

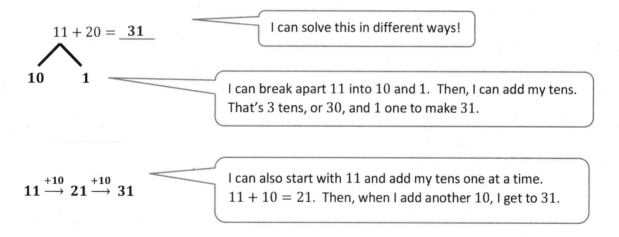

$11 + 20 =$ __31__

I can solve this in different ways!

10   1

I can break apart 11 into 10 and 1. Then, I can add my tens. That's 3 tens, or 30, and 1 one to make 31.

$11 \xrightarrow{+10} 21 \xrightarrow{+10} 31$

I can also start with 11 and add my tens one at a time. $11 + 10 = 21$. Then, when I add another 10, I get to 31.

Name _____ Date _____

Draw quick tens and ones to help you solve the addition problems.

1. 17 + 2 = _____	2. 17 + 3 = _____
3. 14 + 3 = _____	4. 24 + 10 = _____

Make a number bond or use the arrow way to solve the addition problems.

5. 6 + 24 = _____	6. 14 + 20 = _____

7. Solve each addition sentence, and match.

a.

22 + 1 = _____

b.

13 + 6 = _____

c.

3 + 26 = _____

d.

37 + 3 = _____

```
        +3
26 ————————→ 29
```

e.

22 + 10 = _____

```
    13 + 6
     ∧
   10   3
```

EUREKA
MATH

1. Use quick ten drawings or number bonds to make true number sentences.

    a. $13 + 10 =$ __23__

                                               b. $25 + 5 =$ __30__

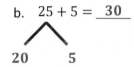

                                                   20   **5**

                                                $5 + 5 = 10$

                                                $10 + 20 = 30$

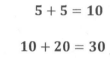

I can draw 13 and then just add another quick ten. Let me count what I have now: 10, 20, …, 23.

I can break apart 25 into 20 and 5. I add 5 and 5 to make the next ten. The next ten is 30.

2. How did you solve Problem 1(a)? Why did you choose to solve it that way?

   I chose to use a quick ten drawing because I only had to draw 1 more ten. That was a fast way to show $13 + 10 = 23.$

3. How did you solve Problem 1(b)? Why did you choose to solve it that way?

   I used a number bond because I wanted to see the parts I had. When I broke apart 25 *into* 20 *and* $5,$ *I saw that I could add* 5 *and* 5 *to make a new ten.*

Name _____   Date _____

Use quick ten drawings or number bonds to make true number sentences.

1.　　　　13 + 20 = _____	2.　　　　23 + 6 = _____
3.　　　　10 + 23 = _____	4.　　　　28 + 6 = _____
5.　　　　26 + 7 = _____	6.　　　　20 + 17 = _____

7. How did you solve Problem 5? Why did you choose to solve it that way?

EUREKA
MATH

Solve using quick ten drawings or number bonds.

8. $23 + 9 = $ _____	9. $27 + 7 = $ _____
10. $24 + 10 = $ _____	11. $20 + 18 = $ _____
12. $28 + 9 = $ _____	13. $29 + 9 = $ _____

14. How did you solve Problem 11? Why did you choose to solve it that way?

Lesson 17: Add ones and ones or tens and tens.

EUREKA MATH

1. Two students both solved the addition problem below using different methods. Are they both correct? Why or why not?

   28 + 5 = __33__

   28 $\xrightarrow{+2}$ 30 $\xrightarrow{+3}$ 33

   This student used the arrow way to get the answer. He used 2 to get to 30 and then added 3 more to get to 33. That means he added 5 altogether to get to 33. That's correct.

   28 + 5 = __33__

   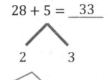
   2   3

   This student broke apart 5 so she could get to the next 10. She needed 2 to get to 30. Then she added the rest and got to 33. That's correct.

   They are both correct. 28 plus 5 is 33. The first student used the arrow way to show his thinking. That student added 2 to get to 30 and then added 3 more since he had to add 5 altogether. The second student used a number bond to show how she got to 33.

2. Another two students solved the same problem shown below, using quick tens. Are they both correct? Why or why not?

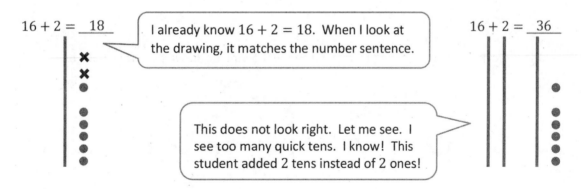

   16 + 2 = __18__

   I already know 16 + 2 = 18. When I look at the drawing, it matches the number sentence.

   This does not look right. Let me see. I see too many quick tens. I know! This student added 2 tens instead of 2 ones!

   16 + 2 = __36__

   The first student is correct. The second student is not correct. The second student added quick tens instead of ones. He has too much.

EUREKA MATH

3. Circle any student work that is correct.

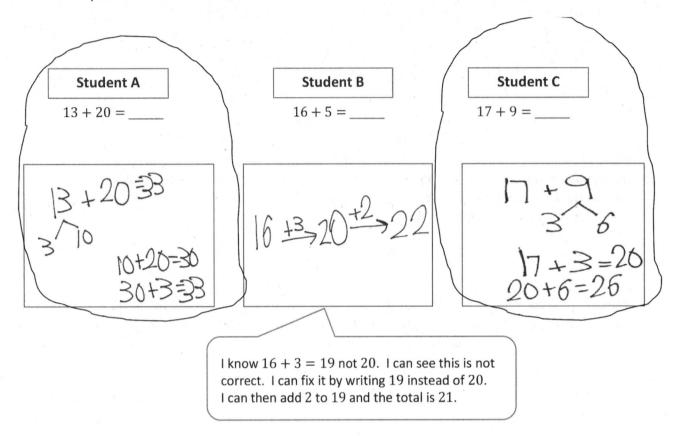

Student A

13 + 20 = ____

13 + 20 = 33
3 ⌐ 10
10+20=30
30+3=33

Student B

16 + 5 = ____

16 +3→ 20 +2→ 22

I know 16 + 3 = 19 not 20. I can see this is not correct. I can fix it by writing 19 instead of 20. I can then add 2 to 19 and the total is 21.

Student C

17 + 9 = ____

17 + 9
3 ⌐ 6
17 + 3 = 20
20+6=26

Fix the student work that was incorrect by making a new drawing or drawings in the space below.

$16 \xrightarrow{+3} 19 \xrightarrow{+2} 21$

Choose a correct student work, and give a suggestion for improvement.

Student A's work can be solved without breaking apart 13. *I can just add* 2 *tens to* 13. *I can do this in my head and get the answer* 33.

Lesson 18:   Share and critique peer strategies for adding two-digit numbers.

EUREKA MATH

Name _____   Date _____

1. Two students both solved the addition problem below using different methods.

18 + 9

18 + 9 = 27
       ⌢
     2   7

18 + 2 = 20
20 + 7 = 27

18 + 9 = 27

18 →⁺² 20 →⁺⁷ 27

18 + 2 = 20
20 + 7 = 27

Are they both correct? Why or why not?

2. Another two students solved the same problem using quick tens.

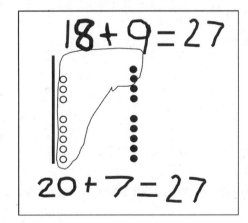

18 + 9 = 29

20 + 9 = 29

18 + 9 = 27

20 + 7 = 27

Are they both correct? Why or why not?

3. Circle any student work that is correct.

19 + 6

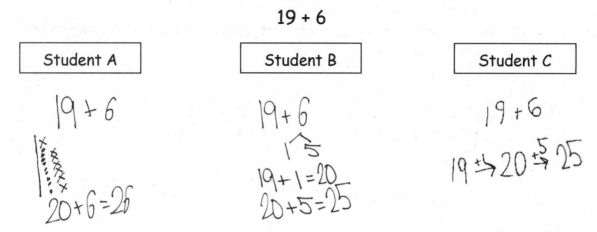

| Student A | Student B | Student C |

Fix the student work that was incorrect by making a new drawing or drawings in the space below.

Choose a correct student work, and give a suggestion for improvement.

76   Lesson 18:   Share and critique peer strategies for adding two-digit numbers.

© 2018 Great Minds®. eureka-math.org

EUREKA MATH®

Solve using the RDW process.

John has 5 red racecars and 12 blue racecars. How many racecars does John have in all?

I can draw 5 circles for the red racecars. I put my circles in a rectangle to keep them organized. I label my drawing with the number 5 and the letter R, so I know that this rectangle represents the 5 red racecars.

I connect the two rectangles and draw a box with a question mark labeled with the letter T because it is the total. When I find the total, I will know the answer to the question.

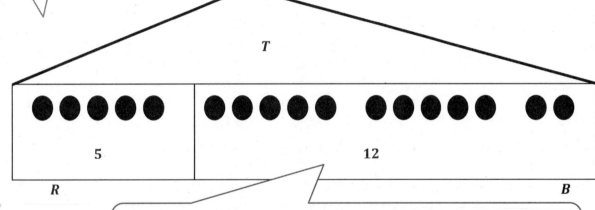

I can draw 12 circles for the blue racecars. I organize my circles and put them in a rectangle labeled with the number 12 and the letter B, so I know that this rectangle represents the 12 blue racecars.

$5 + 12 =$ ☐ 17

I draw a box around 17 because it is the total and answers the question. The last part of RDW is write. I can write a statement to answer the question.

John has 17 racecars.

Lesson 19:   Use tape diagrams as representations to solve *put together/take apart with total unknown* and *add to with result unknown* word problems.

77

EUREKA
MATH®

© 2018 Great Minds®. eureka-math.org

Name _____   Date _____

Read the word problem.
Draw a tape diagram and label.
Write a number sentence and a statement that matches
the story.

1. Darnel is playing with his 4 red robots. Ben joins him with 13 blue robots.
   How many robots do they have altogether?

They have _____ robots.

2. Rose and Emi had a jump rope contest. Rose jumped 14 times, and Emi jumped
   6 times. How many times did Rose and Emi jump?

They jumped _____ times.

Lesson 19:   Use tape diagrams as representations to solve *put together/take apart*
                with total unknown and *add to with result unknown* word problems.

© 2018 Great Minds®. eureka-math.org

79

3. Pedro counted the airplanes taking off and landing at the airport. He saw 7 airplanes take off and 6 airplanes land. How many airplanes did he count altogether?

Pedro counted _____ airplanes.

4. Tamra and Willie scored all the points for their team in their basketball game. Tamra scored 13 points, and Willie scored 5 points. What was their team's score for the game?

The team's score was _____ points.

Lesson 19:   Use tape diagrams as representations to solve *put together/take apart with total unknown* and *add to with result unknown* word problems.

© 2018 Great Minds®. eureka-math.org

EUREKA MATH

What can I draw?

Solve using the RDW process.

1. Mary has 14 play practices this month. 7 practices are after school, and the rest are in the evening.
    How many practices are in the evening?

What do I know after reading the problem?

I know the total, or the whole. I can draw 14 circles in 5-group rows to represent the total number of practices.

T

14

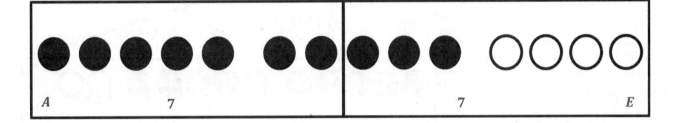

I know there are 7 practices after school. I can draw a rectangle around 7 of the circles to represent the 7 practices that are after school. I label the rectangle with the letter *A* for after school.

I draw a rectangle around the rest of the circles. This represents the practices that are in the evening. I count the circles and see there are 7 practices in the evening. I label the rectangle with the letter *E* for evening.

$14 - 7 = \boxed{7}$

I draw a rectangle around the 7 because 7 is the answer to the question.

Mary has 7 practices in the evening.

EUREKA MATH®

Lesson 20:   Recognize and make use of part–whole relationships within tape
               diagrams when solving a variety of problem types.

81

© 2018 Great Minds®. eureka-math.org

2. Katelyn gave some of her stickers to her friend. She had 18 stickers at first, and she still has 12 stickers left. How many stickers did Katelyn give to her friend?

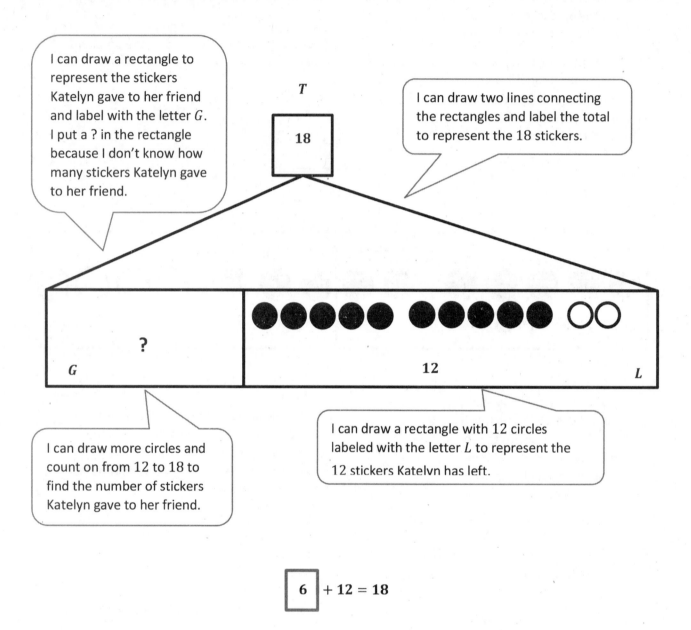

I can draw a rectangle to represent the stickers Katelyn gave to her friend and label with the letter G. I put a ? in the rectangle because I don't know how many stickers Katelyn gave to her friend.

T

18

I can draw two lines connecting the rectangles and label the total to represent the 18 stickers.

?

G

12

L

I can draw more circles and count on from 12 to 18 to find the number of stickers Katelyn gave to her friend.

I can draw a rectangle with 12 circles labeled with the letter L to represent the 12 stickers Katelyn has left.

$$6 + 12 = 18$$

Katelyn gave 6 stickers to her friend.

Lesson 20: Recognize and make use of part–whole relationships within tape diagrams when solving a variety of problem types.

© 2018 Great Minds®. eureka-math.org

EUREKA MATH

Name _____ Date _____

Read the word problem.
Draw a tape diagram and label.
Write a number sentence and a statement that matches
the story.

1. Rose has 12 soccer practices this month. 6 practices are in the afternoon, but the
    rest are in the morning. How many practices will be in the morning?

Rose has _____ practices in the morning.

2. Ben caught 16 fish. He put some back in the lake. He brought home 7 fish.
    How many fish did he put back in the lake?

Ben put _____ fish back in the lake.

Lesson 20: Recognize and make use of part–whole relationships within tape
diagrams when solving a variety of problem types.

© 2018 Great Minds®. eureka-math.org

83

3. Nikil solved 9 problems on the first Sprint. He solved 11 problems on the second Sprint. How many problems did he solve on the two Sprints?

Nikil solved _____ problems on the Sprints.

4. Shanika returned some books to the library. She had 16 books at first, and she still has 13 books left. How many books did she return to the library?

Shanika returned _____ books to the library.

Lesson 20:   Recognize and make use of part–whole relationships within tape diagrams when solving a variety of problem types.

EUREKA MATH

Solve using the RDW process.

Emi made a bracelet that was 13 centimeters long. The bracelet didn't fit so she made the bracelet longer.
Now the bracelet is 17 centimeters long. How many centimeters did Emi add to the bracelet?

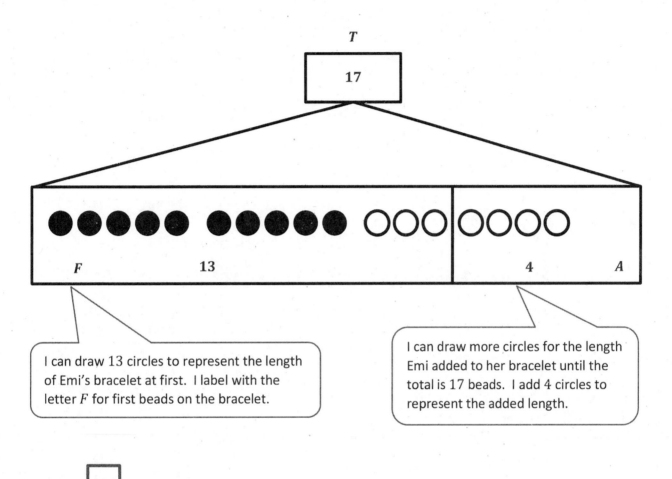

T

17

F 13   4 A

I can draw 13 circles to represent the length of Emi's bracelet at first. I label with the letter F for first beads on the bracelet.

I can draw more circles for the length Emi added to her bracelet until the total is 17 beads. I add 4 circles to represent the added length.

$$13 + \boxed{4} = 17$$

Emi added 4 centimeters to the bracelet.

EUREKA
MATH

Lesson 21:   Recognize and make use of part–whole relationships within tape
             diagrams when solving a variety of problem types.

85

© 2018 Great Minds®. eureka-math.org

Name _____   Date _____

Read the word problem.
Draw a tape diagram and label.
Write a number sentence and a statement that matches
the story.

1. Fatima has 12 colored pencils in her bag. She has 6 regular pencils, too. How many
    pencils does Fatima have?

                                          Fatima has _____ pencils.

2. Julio swam 7 laps in the morning. In the afternoon, he swam some more laps.
    He swam a total of 14 laps. How many laps did he swim in the afternoon?

                                          Julio swam _____ laps in the afternoon.

3. Peter built 18 models. He built 13 airplanes and some cars. How many car models
    did he build?

                                          Peter built _____ car models.

Lesson 21:   Recognize and make use of part–whole relationships within tape
              diagrams when solving a variety of problem types.

© 2018 Great Minds®. eureka-math.org

87

4. Kiana found some shells at the beach. She gave 8 shells to her brother. Now, she has 9 shells left. How many shells did Kiana find at the beach?

Kiana found _____ shells.

Lesson 21:   Recognize and make use of part–whole relationships within tape diagrams when solving a variety of problem types.

© 2018 Great Minds®. eureka-math.org

EUREKA MATH®

Use the tape diagrams to write a variety of word problems. Use the word bank, if needed. Remember to label your model after you write the story.

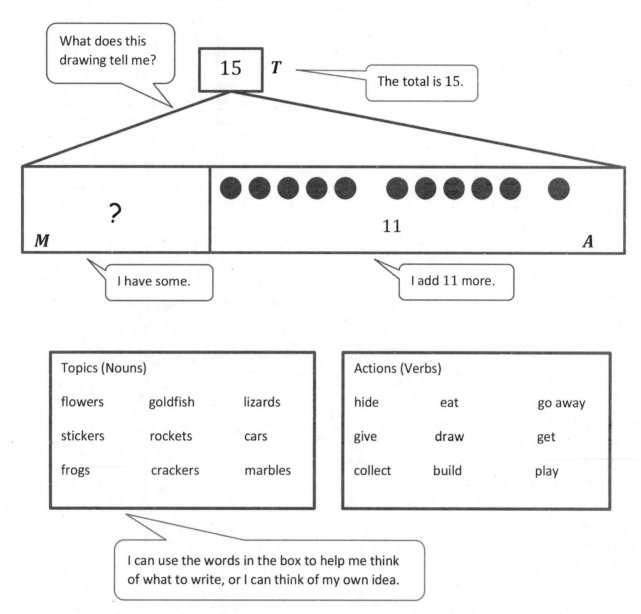

Beth picks *some* **flowers for her mom in the morning. She picks** 11 *more flowers in the afternoon. Now she* **has** 15 *flowers for her mom. How many flowers did Beth pick in the morning?*

Name _____   Date _____

Use the tape diagrams to write a variety of word problems. Use the word bank if needed. Remember to label your model after you write the story.

Topics (Nouns)		
flowers	goldfish	lizards
stickers	rockets	cars
frogs	crackers	marbles

Actions (Verbs)		
hide	eat	go away
give	draw	get
collect	build	play

1.

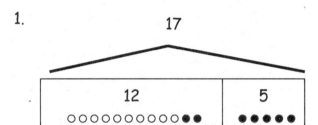

17

12   5

2.

16

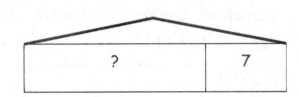

?	7

EUREKA
MATH®

1. Fill in the blanks, and match the pairs that show the same amount.

I can match these pictures because they both show 32. 3 tens 2 ones is equal to 2 tens 12 ones. If I bundle 10 ones in the picture on the right, it would have 3 tens 2 ones.

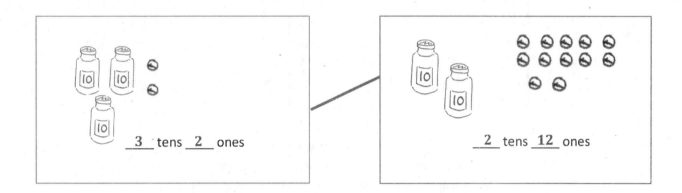

__3__ tens __2__ ones

__2__ tens __12__ ones

2. Match the place value charts that show the same amount.

The place value chart shows how many tens and ones. It's okay to have more than 9 in the ones. 2 tens 15 ones is 35.

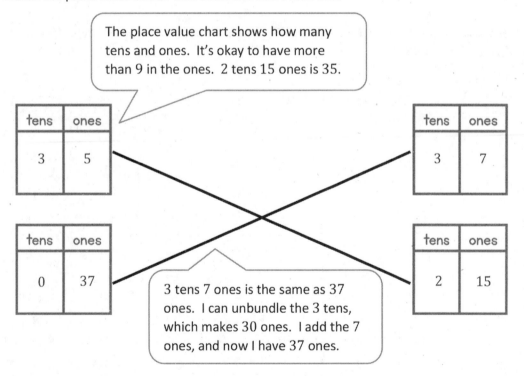

3 tens 7 ones is the same as 37 ones. I can unbundle the 3 tens, which makes 30 ones. I add the 7 ones, and now I have 37 ones.

Lesson 23:   Interpret two-digit numbers as tens and ones, including cases with more than 9 ones.

93

EUREKA MATH

3. Emi says 29 is the same as 1 ten 19 ones, and Ben says 29 is the same as 2 tens 19 ones. Draw quick tens to show if Emi or Ben is correct.

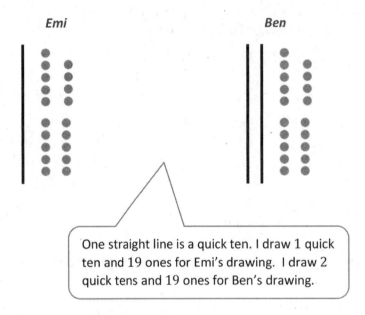

Emi   Ben

> One straight line is a quick ten. I draw 1 quick ten and 19 ones for Emi's drawing. I draw 2 quick tens and 19 ones for Ben's drawing.

Emi is correct because 1 ten 19 ones is the same as 29. Ben is not correct because 2 tens 19 ones is the same as 39, which is not 29.

Lesson 23: Interpret two-digit numbers as tens and ones, including cases with more than 9 ones.

EUREKA
MATH®

Name _____   Date _____

1. Fill in the blanks, and match the pairs that show the same amount.

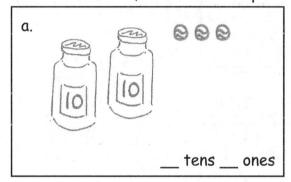

a.

___ tens ___ ones

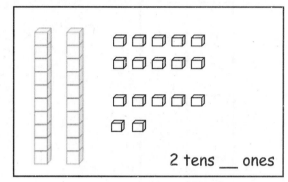

2 tens ___ ones

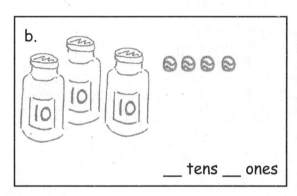

b.

___ tens ___ ones

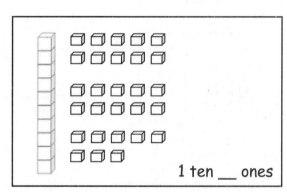

1 ten ___ ones

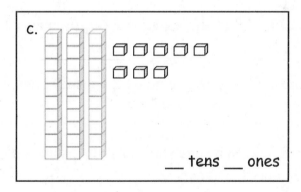

c.

___ tens ___ ones

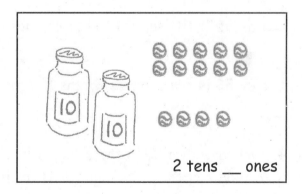

2 tens ___ ones

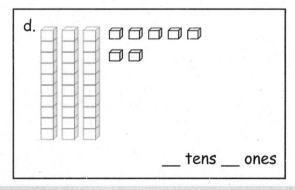

d.

___ tens ___ ones

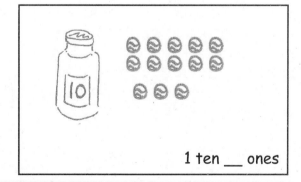

1 ten ___ ones

Lesson 23:   Interpret two-digit numbers as tens and ones, including cases with
more than 9 ones.

© 2018 Great Minds®. eureka-math.org   95

2. Match the place value charts that show the same amount.

a.
tens	ones
2	18

tens	ones
3	8

b.
tens	ones
1	16

tens	ones
2	1

c.
tens	ones
0	21

tens	ones
2	6

3. Check each sentence that is true.

☐ a. 35 is the same as 1 ten 25 ones.   ☐ b. 28 is the same as 1 ten 18 ones.

☐ c. 36 is the same as 2 tens 16 ones.   ☐ d. 39 is the same as 2 tens 29 ones.

4. Emi says that 37 is the same as 1 ten 27 ones, and Ben says that 37 is the same as 2 tens 7 ones. Draw quick tens to show if Emi or Ben is correct.

Lesson 23:   Interpret two-digit numbers as tens and ones, including cases with more than 9 ones.

© 2018 Great Minds®. eureka-math.org

EUREKA
MATH

1. Solve using number bonds. Write the two number sentences that show that you added 10 first. Draw quick tens and ones if that helps you.

a.

$15 + 13 =$ __28__

10   3

$15 + 10 = 25$

$25 + 3 = 28$

b.

$16 + 23 =$ __39__

10   6

$23 + 10 =$ __33__

__33__ $+ 6 =$ __39__

> I draw 15 using quick tens and ones. I can break apart 13 into 10 and 3. I add 15 and 10, which equals 25. I add the 3 ones to 25. I use x's to show I am adding the 3 ones.

> I want to add 10 first, so I break apart 16 into 10 and 6 using a number bond. I add 10 to 23 and get 33. Then, I add 33 and 6, which is my answer of 39.

2. Solve using number bonds.

a.

$17 + 23 =$ __40__

10   7

$23 + 10 = 33$

$33 + 7 = 40$

b.

$22 + 18 =$ __40__

10   8

> I can break apart 17 into 10 and 7 using a number bond. I add 10 and 23, which equals 33. Then, I add 33 and 7 to get my answer of 40.

> I didn't write the two number sentences because I was able to add in my head.

Lesson 24: Add a pair of two-digit numbers when the ones digits have a sum less than or equal to 10.

97

© 2018 Great Minds®. eureka-math.org

Name _____   Date _____

1. Solve using number bonds. Write the two number sentences that show that you added the ten first. Draw quick tens and ones if that helps you.

---

a.

$13 + 16 =$ _____

10   3

$16 + 10 = 26$

$26 + 3 = 29$

---

b.

$16 + 23 =$ _____

10   6

$23 + 10 =$ _____

_____ $+ 6 =$ _____

---

c.

$16 + 14 =$ _____

10   4

$16 + 10 =$ _____

_____ $+ 4 =$ _____

---

d.

$14 + 26 =$ _____

10   4

$26 + 10 =$ _____

_____ $+$ _____ $=$ _____

---

e.

$17 + 13 =$ _____

10   3

____ $+$ ____ $=$ _____

____ $+$ ____ $=$ _____

---

f.

$27 + 13 =$ _____

____ $+$ ____ $=$ _____

____ $+$ ____ $=$ _____

Lesson 24:   Add a pair of two-digit numbers when the ones digits have a sum less than or equal to 10.

© 2018 Great Minds®. eureka-math.org

99

2. Solve using number bonds. Part (a) has been started for you.

a.	b.
$14 + 13 = \underline{\hspace{1cm}}$ 10 3 $\underline{\hspace{1cm}} + \underline{\hspace{1cm}} = \underline{\hspace{1cm}}$ $\underline{\hspace{1cm}} + \underline{\hspace{1cm}} = \underline{\hspace{1cm}}$	$24 + 14 = \underline{\hspace{1cm}}$ $\underline{\hspace{1cm}} + \underline{\hspace{1cm}} = \underline{\hspace{1cm}}$ $\underline{\hspace{1cm}} + \underline{\hspace{1cm}} = \underline{\hspace{1cm}}$
c. $15 + 14 = \underline{\hspace{1cm}}$	d. $24 + 15 = \underline{\hspace{1cm}}$
e. $22 + 17 = \underline{\hspace{1cm}}$	f. $27 + 12 = \underline{\hspace{1cm}}$
g. $18 + 12 = \underline{\hspace{1cm}}$	h. $28 + 12 = \underline{\hspace{1cm}}$

Lesson 24:   Add a pair of two-digit numbers when the ones digits have a sum less than or equal to 10.

© 2018 Great Minds®. eureka-math.org

EUREKA MATH

1. Solve using number bonds. This time, add the tens first. Write the two number sentences to show what you did.

   a.

   $12 + 16 = \underline{\textbf{28}}$

   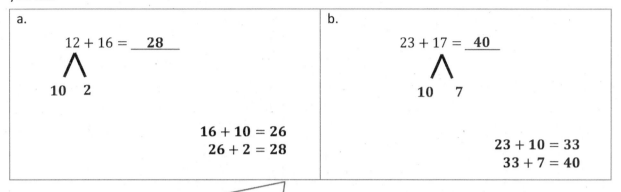

   10   2

   $16 + 10 = 26$
   $26 + 2 = 28$

   b.

   $23 + 17 = \underline{\textbf{40}}$

   10   7

   $23 + 10 = 33$
   $33 + 7 = 40$

   > I need to add the tens first. I can break apart 12 into 10 and 2 and add 10 to 16 first. $10 + 16 = 26$. I still have 2 more to add: $26 + 2 = 28$.

2. Solve using number bonds. This time, add the ones first. Write the two number sentences to show what you did.

   a.

   $23 + 16 = \underline{\textbf{39}}$

   6   10

   $23 + 6 = 29$
   $29 + 10 = 39$

   b.

   $11 + 29 = \underline{\textbf{40}}$

   10   1

   $29 + 1 = 30$
   $30 + 10 = 40$

   > I can still break apart 16 into 6 and 10, but this time I add the 6 ones to 23 first.

   > I notice that when I add my ones, the result is the next 10.

EUREKA MATH

Lesson 25:   Add a pair of two-digit numbers when the ones digits have a sum less than or equal to 10.

101

© 2018 Great Minds®. eureka-math.org

Name _____ Date _____

1. Solve using number bonds. This time, add the tens first. Write the 2 number sentences to show what you did.

a.	b.
12 + 14 = _____	14 + 21 = _____

c.	d.
15 + 14 = _____	25 + 14 = _____

e.	f.
23 + 16 = _____	16 + 24 = _____

EUREKA MATH®

Lesson 25:   Add a pair of two-digit numbers when the ones digits have a sum less than or equal to 10.

© 2018 Great Minds®. eureka-math.org

103

2. Solve using number bonds. This time, add the ones first. Write the 2 number sentences to show what you did.

a. 27 + 10 = _____	b. 27 + 13 = _____
c. 13 + 26 = _____	d. 26 + 14 = _____
e. 12 + 18 = _____	f. 18 + 21 = _____
g. 19 + 11 = _____	h. 21 + 19 = _____

Lesson 25:   Add a pair of two-digit numbers when the ones digits have a sum less than or equal to 10.

© 2018 Great Minds®. eureka-math.org

EUREKA MATH

1. Solve using a number bond to add ten first. Write the two addition sentences that help you.

> I need to use the add ten first strategy. I break apart one of the numbers into 10 and some ones.

a. $25 + 14 =$ __39__

10   4

$25 + 10 =$ __35__

__35__ + __4__ = __39__

b. $19 + 15 =$ __34__

10   5

$19 + 10 =$ __29__

__29__ + __5__ = __34__

> Adding 10 to a number is easy. I know $25 + 10 = 35$.
> Now I just have to add the ones; that's easy too.

2. Solve using a number bond to make a ten first. Write the two number sentences that help you.

a. $16 + 19 =$ __35__

15   1

__19__ $+ 1 =$ __20__

__20__ $+ 15 =$ __35__

b. $18 + 14 =$ __32__

2   12

__18__ $+$ __2__ $=$ __20__

__20__ $+$ __12__ $=$ __32__

> 16 is broken apart into 15 and 1 because 19 needs 1 more to make the next ten.

> I could have also chosen to break apart 18 into 6 and 12 because I can make the next ten with 6 and 14.

Lesson 26: Add a pair of two-digit numbers when the ones digits have a sum greater than 10.

105

© 2018 Great Minds®. eureka-math.org

Name _____   Date _____

1. Solve using a number bond to add ten first. Write the 2 addition sentences that helped you.

a.
18 + 13 = _____

10   3

18 + 10 = 28

28 + 3 = 31

b.
13 + 19 = _____

10   3

19 + 10 = 29

29 + 3 = 32

c.
17 + 15 = _____

10   5

17 + 10 = _____

_____ + 5 = _____

d.
17 + 16 = _____

10   6

17 + 10 = _____

_____ + 6 = _____

e.
17 + 14 = _____

10   4

17 + 10 = _____

_____ + _____ = _____

f.
19 + 17 = _____

10   7

19 + 10 = _____

_____ + _____ = _____

EUREKA MATH

Lesson 26:   Add a pair of two-digit numbers when the ones digits have a sum greater than 10.

107

© 2018 Great Minds®. eureka-math.org

2. Solve using a number bond to make a ten first. Write the 2 number sentences that helped you.

a.
19 + 13 = _____

1   12

19 + 1 = 20

20 + 12 = 32

b.
19 + 14 = _____
1   13

19 + 1 = 20

20 + 13 = 33

c.
18 + 15 = _____
2   13

18 + 2 = _____

20 + 13 = _____

d.
18 + 17 = _____
2   15

18 + 2 = _____

_____ + 15 = _____

e.
18 + 19 = _____

17   1

_____ + 1 = _____

_____ + 17 = _____

f.
19 + 19 = _____

18   1

_____ + _____ = _____

_____ + _____ = _____

Lesson 26:   Add a pair of two-digit numbers when the ones digits have a sum greater than 10.

© 2018 Great Minds®. eureka-math.org

EUREKA
MATH

For the following problems, solve using the strategy that makes you feel most comfortable.

1. 15 + 17 = ___32___

10   5

17 + 10 = 27
27 + 5 = 32

I feel more comfortable using quick tens and ones. I can draw 17 with one quick ten and 7 ones. I draw the ones with 5 closed circles and 2 open circles, to help me see how many more 7 needs to make a new ten.

I can break apart 15 into 10 and 5, and add a quick ten next to the quick ten in 17. Now I only have 5 more to add. I use x's to draw this part to help keep track of how many I need to draw. I add 3 x's to the 7 ones in 17 . I draw a line through the circles and x's because 7 and 3 makes a ten, I have 2 more to draw, I can draw 2 more x's. My drawing shows 32.

2. 18 + 14 = __32__

18 + 10 = 28
28 + 4 = 32

For this problem, I feel most comfortable using the add ten first strategy, which means I break apart 14 into 10 and 4, and then I add 10 and 18 which makes 28. I have 4 more to add. 28 and 4 is 32.

3. 19 + 12 = __31__

19 + 2 = 21
21 + 10 = 31

For this problem, I feel most comfortable adding the ones first. 12 is ten and 2. I can add the 2 to 19, which makes 21. Then, I can quickly add the 10 to get the answer.

4. 19 + 18 = __37__

19 + 1 = 20
20 + 17 = 37

For this problem, I feel most comfortable making a 10. I know that 19 needs one more to make 20. I can easily break apart 18 into 1 and 17.

EUREKA MATH

Lesson 27:   Add a pair of two-digit numbers when the ones digits have a sum greater than 10.

109

© 2018 Great Minds®. eureka-math.org

Name _____ Date _____

1. Solve using number bonds with pairs of number sentences. You may draw quick tens and some ones to help you.

a. 17 + 14 = _____	b. 16 + 15 = _____
c. 17 + 15 = _____	d. 18 + 13 = _____
e. 18 + 15 = _____	f. 18 + 16 = _____
g. 19 + 15 = _____	h. 19 + 16 = _____

EUREKA MATH

Lesson 27:   Add a pair of two-digit numbers when the ones digits have a sum greater than 10.

111

© 2018 Great Minds®. eureka-math.org

2. Solve. You may draw quick tens and some ones to help you.

a. 19 + 14 = _____	b. 19 + 17 = _____
c. 18 + 17 = _____	d. 16 + 16 = _____
e. 17 + 14 = _____	f. 15 + 16 = _____
g. 19 + 19 = _____	h. 18 + 18 = _____

Lesson 27:   Add a pair of two-digit numbers when the ones digits have a sum
               greater than 10.

EUREKA
MATH®

Solve using quick tens and ones, number bonds, or the arrow way.

1. $26 + 13 =$ __39__

$$26 \xrightarrow{+10} 36 \xrightarrow{+3} 39$$

> I solved using the arrow way because I know 13 is 10 and 3. I can add the 10 first to get 36 and then add 3. My answer is 39.

2. $18 + 18 =$ __36__

```
    /\
   2   16
```

$18 + 2 = 20$

$20 + 16 = 36$

> I solved using a number bond. I made a ten. I know 18 needs 2 more to make 20, so I broke apart the other 18 into 2 and 16. I added 20 and 16 to get my answer of 36.

3. $22 + 18 =$ __40__

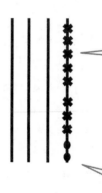

> I solved using quick tens and ones. I can draw 2 quick tens and 2 ones. I can draw 18 more. 18 is 1 ten and 8 ones.

> I can draw the 2 ones in 22 with circles and the 8 ones in 18 with x's. When I do this I make a new ten and draw a line through it.

EUREKA MATH®

Lesson 28:   Add a pair of two-digit numbers with varied sums in the ones.

113

Name _____ Date _____

Solve using quick tens and ones, number bonds, or the arrow way.

a. 13 + 16 = _____	b. 15 + 16 = _____
c. 16 + 16 = _____	d. 26 + 12 = _____
e. 22 + 17 = _____	f. 17 + 15 = _____
g. 17 + 16 = _____	h. 18 + 17 = _____

EUREKA MATH

Lesson 28: Add a pair of two-digit numbers with varied sums in the ones.

115

© 2018 Great Minds®. eureka-math.org

i. 24 + 13 = _____

j. 15 + 24 = _____

k. 19 + 16 = _____

l. 14 + 22 = _____

m. 27 + 12 = _____

n. 28 + 12 = _____

o. 18 + 17 = _____

p. 19 + 18 = _____

Lesson 28:   Add a pair of two-digit numbers with varied sums in the ones.

EUREKA MATH

Solve using quick tens and ones, number bonds, or the arrow way.

1. $24 + 16 =$ ___**40**___

    $24 \xrightarrow{+10} 34 \xrightarrow{+6} 40$

    > I solved using the arrow way because I know 16 is 10 and 6. I can add the 10 to 24 first to get 34. I know that 34 and 6 is 40.

2. $17 + 12 =$ ___**29**___

    $\bigwedge$
    10   **2**

    > I solved using a number bond. I added 17 and 10 and got 27. Then I added 27 and 2 to get my answer of 29. I didn't need to write the number sentences because I can do the math in my head.

    > I didn't solve any using drawings this time. Using the arrow way and number bonds is more efficient for me now. If I get stuck I can always use a quick ten drawing.

EUREKA MATH

Lesson 29:   Add a pair of two-digit numbers with varied sums in the ones.

117

© 2018 Great Minds®. eureka-math.org

Name _____   Date _____

1. Solve using quick ten drawings, number bonds, or the arrow way.

a. 13 + 15 = _____	b. 26 + 12 = _____
c. 23 + 16 = _____	d. 17 + 16 = _____
e. 14 + 17 = _____	f. 27 + 12 = _____
g. 15 + 18 = _____	h. 18 + 16 = _____

Lesson 29:   Add a pair of two-digit numbers with varied sums in the ones.

© 2018 Great Minds®. eureka-math.org

119

2. Solve using quick ten drawings, number bonds, or the arrow way.

a. $17 + 12 = \_\_\_\_$	b. $21 + 17 = \_\_\_\_$
c. $17 + 15 = \_\_\_\_$	d. $27 + 13 = \_\_\_\_$
e. $23 + 14 = \_\_\_\_$	f. $18 + 17 = \_\_\_\_$
g. $18 + 11 = \_\_\_\_$	h. $18 + 18 = \_\_\_\_$

Lesson 29:   Add a pair of two-digit numbers with varied sums in the ones.

EUREKA
MATH

Grade 1
Module 5

1. Circle the shapes that have exactly 3 corners.

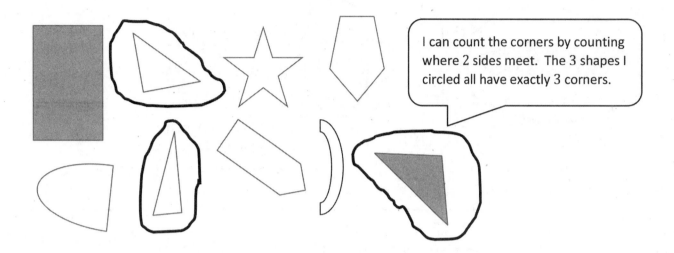

I can count the corners by counting where 2 sides meet. The 3 shapes I circled all have exactly 3 corners.

2. Circle the shapes that have no square corners.

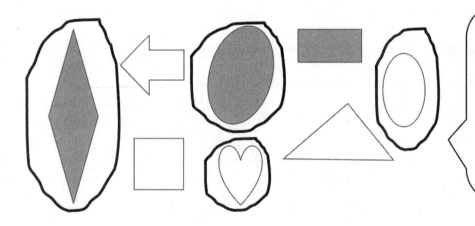

I can use my square corner tester, a paper shaped like an "L", to see if these shapes have square corners. I put the corner of the tester in the corner of the shape. If the corners match, the shape has square corners.

Lesson 1:   Classify shapes based on defining attributes using examples, variants, and non-examples.

123

3. Circle the shapes that have no straight sides.

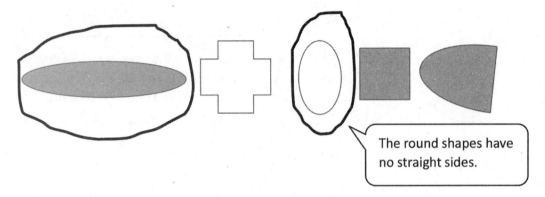

The round shapes have no straight sides.

4.

a. Draw a shape that has only square corners.	b. Draw another shape with only square corners that is different from the shape you drew in part (a) and from the ones above.
	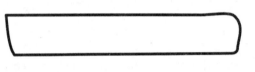

5. Which attributes, or characteristics, are the same for all of the shapes in Group A?
    GROUP A

   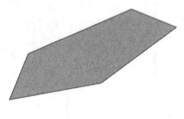

They all ___*have 5 straight sides*___.

They all ___*have 5 corners*___.

Lesson 1:   Classify shapes based on defining attributes using examples, variants, and non-examples.

EUREKA MATH

6.

a. Circle the shape that best fits with Group A in Problem 5.

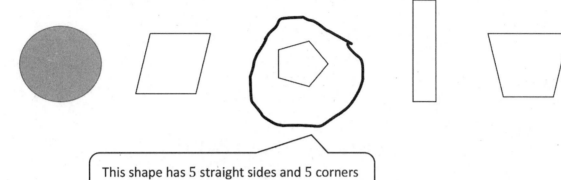

This shape has 5 straight sides and 5 corners just like the shapes from Group A!

b. Draw 2 more shapes that would fit with Group A.

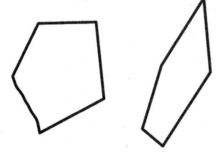

c. Draw 1 shape that would **not** fit with Group A.

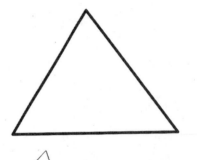

I can draw any shape I want, as long as it doesn't have 5 straight sides and 5 corners!

Lesson 1: Classify shapes based on defining attributes using examples, variants, and non-examples.

125

© 2018 Great Minds®. eureka-math.org

Name _____   Date _____

1. Circle the shapes that have 3 straight sides.

2. Circle the shapes that have no corners.

3. Circle the shapes that have only square corners.

4.

| a. Draw a shape that has 4 straight sides. | b. Draw another shape with 4 straight sides that is different from 4(a) and from the ones above. |

EUREKA MATH

Lesson 1:   Classify shapes based on defining attributes using examples, variants, and non-examples.

© 2018 Great Minds®. eureka-math.org

127

5. Which attributes, or characteristics, are the same for all of the shapes in Group A?

GROUP A

They all _____.

They all _____.

6. Circle the shape that best fits with Group A.

7. Draw 2 more shapes that would fit in Group A.	8. Draw 1 shape that would **not** fit in Group A.

Lesson 1:   Classify shapes based on defining attributes using examples, variants, and non-examples.

EUREKA MATH

1. Color the shapes using the key. Write the number of shapes you colored on each line.

Key

RED—4 straight sides: __8__

GREEN—3 straight sides: __8__

BLUE—6 straight sides: __2__

YELLOW—0 straight sides: __3__

I count each side to know which color to make it. I know that yellow will be a circle because round shapes have no straight sides!

A triangle has __3__ straight sides and __3__ corners.

I colored __8__ triangles.

A hexagon has __6__ straight sides and __6__ corners.

I colored __2__ hexagons.

A circle has __0__ straight sides and __0__ corners.

I colored __3__ circles.

A rhombus has __4__ straight sides that are equal in length and __4__ corners.

I colored __3__ rhombuses.

The cat's neck and body look like squares. Squares are rhombuses, too! The cat's tie also is a rhombus. That makes 3 rhombuses.

EUREKA MATH

Lesson 2:   Find and name two-dimensional shapes including trapezoid, rhombus,
              and a square as a special rectangle, based on defining attributes of
              sides and corners.
© 2018 Great Minds®. eureka-math.org

129

2. A triangle is a closed shape with 3 straight sides and 3 corners.

a. Cross off the shape that is **not** a triangle.

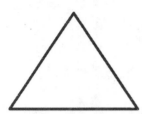  

b. Explain your thinking: _**The shape that I crossed off is not a triangle because it is missing**_

**an open shape and doesn't have 3 sides.**

Lesson 2: Find and name two-dimensional shapes including trapezoid, rhombus, and a square as a special rectangle, based on defining attributes of sides and corners.

EUREKA
MATH

Name _____   Date _____

1. Color the shapes using the key. Write the number of shapes you colored on each line.

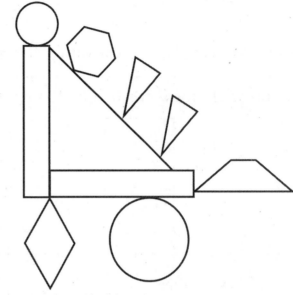

Key

RED 3 straight sides: _____

BLUE 4 straight sides: _____

GREEN 6 straight sides: _____

YELLOW 0 straight sides: _____

2.

   a. A **triangle** has ____ straight sides and ____ corners.

   b. I colored ____ triangles.

3.

   a. A **hexagon** has ____ straight sides and ____ corners.

   b. I colored ____ hexagon.

4.

   a. A **circle** has ____ straight sides and ____ corners.

   b. I colored ____ circles.

5.

    a. A **rhombus** has ____ straight sides that are equal in length and ____ corners.

    b. I colored ____ rhombus.

6. A **rectangle** is a closed shape with 4 straight sides and 4 square corners.

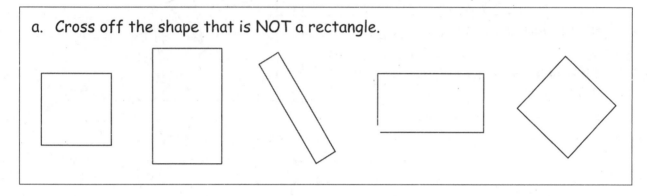

    b. Explain your thinking: _____

    _____

7. A **rhombus** is a closed shape with 4 straight sides of the same length.

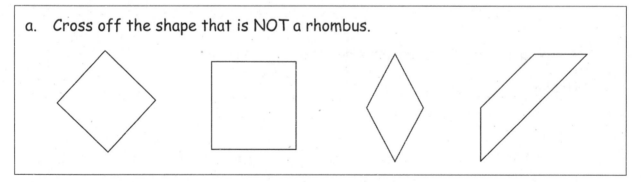

    b. Explain your thinking: _____

    _____

Lesson 2:   Find and name two-dimensional shapes including trapezoid, rhombus, and a square as a special rectangle, based on defining attributes of sides and corners.
© 2018 Great Minds®. eureka-math.org

1. Go on a scavenger hunt for 3-dimensional shapes. Look for objects that would fit in the chart below.

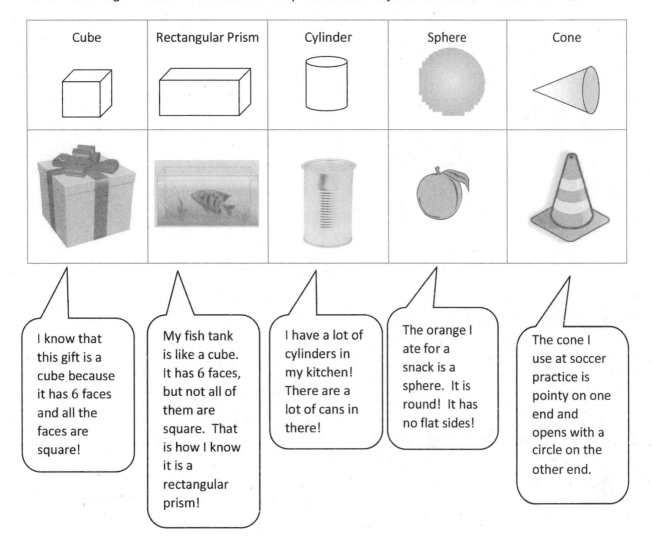

Cube	Rectangular Prism	Cylinder	Sphere	Cone

I know that this gift is a cube because it has 6 faces and all the faces are square!

My fish tank is like a cube. It has 6 faces, but not all of them are square. That is how I know it is a rectangular prism!

I have a lot of cylinders in my kitchen! There are a lot of cans in there!

The orange I ate for a snack is a sphere. It is round! It has no flat sides!

The cone I use at soccer practice is pointy on one end and opens with a circle on the other end.

EUREKA MATH

Lesson 3: Find and name three-dimensional shapes including cone and rectangular prism, based on defining attributes of faces and points.

© 2018 Great Minds®. eureka-math.org

133

Name _____   Date _____

1. Go on a scavenger hunt for 3-dimensional shapes. Look for objects at home that would fit in the chart below. Try to find at least four objects for each shape.

Cube	Rectangular Prism	Cylinder	Sphere	Cone

EUREKA MATH

Lesson 3:   Find and name three-dimensional shapes including cone and rectangular prism, based on defining attributes of faces and points.

© 2018 Great Minds®. eureka-math.org

135

2. Choose one object from each column. Explain how you know that object belongs in that column. Use the word bank if needed.

Word Bank

faces	circle	square	roll	six
sides	rectangle	point	flat	

a. I put the _____ in the cube column because

_____ .

b. I put the _____ in the cylinder column because

_____ .

c. I put the _____ in the sphere column because

_____ .

d. I put the _____ in the cone column because

_____ .

e. I put the _____ in the rectangular prism column

because _____ .

Lesson 3:   Find and name three-dimensional shapes including cone and
rectangular prism, based on defining attributes of faces and points.

EUREKA
MATH

1. Cut out the pattern block shapes from the bottom of the page. Color them to match the key, which is different from the pattern block colors in class. Trace or draw to show what you did.

| Hexagon—purple | Triangle—orange | Rhombus—pink | Trapezoid—brown |

Use 3 rhombuses to make a hexagon.

Pink   Pink

Pink

Use 1 trapezoid, 1 rhombus, and 1 triangle to make 1 hexagon.

Pink

Orange

Brown

I can make a bigger shape, or a composite shape, by putting smaller shapes together!

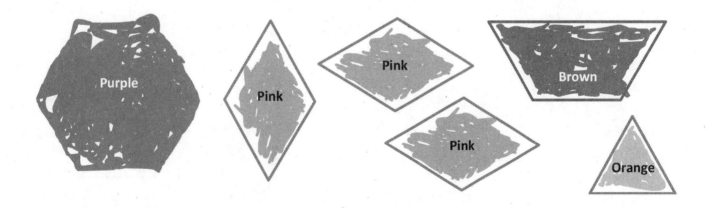

Purple

Pink

Pink

Pink

Brown

Pink

Orange

2. How many smaller squares do you see in this square?

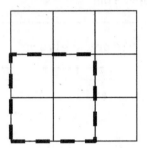

I can find ___**13**___ squares in this large square.

I know each little individual square counts as 1, so that makes 9. There are also 4 medium squares that are made of 4 little squares, so altogether that makes 13.

Name _____   Date _____

Cut out the pattern block shapes from the bottom of the page. Color them to match the key, which is different from the pattern block colors in class. Trace or draw to show what you did.

Hexagon—red	Triangle—blue	Rhombus—yellow	Trapezoid—green

1. Use 3 triangles to make 1 trapezoid.	2. Use 3 triangles to make 1 trapezoid, and then add 1 trapezoid to make 1 hexagon.

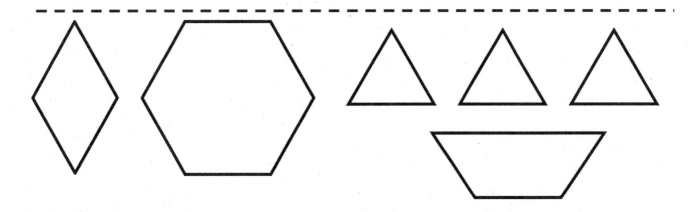

3. How many squares do you see in this large square?

I can find _____ squares in this rectangle.

Lesson 4:   Create composite shapes from two-dimensional shapes.

© 2018 Great Minds®. eureka-math.org

Use your tangram pieces to complete the problems below.

Draw or trace to show the parts you used to make the shape.

1. Use 2 triangles to make a square.

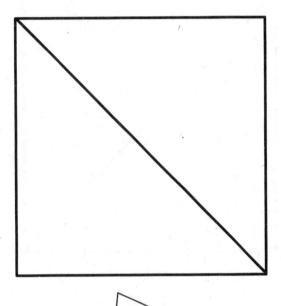

I can make a square with two triangles just like I did in class! I know that if I fold a square in half diagonally, it will make two triangles, so I just put my triangles together with the long sides touching, and it makes a square!

2. Use the square you made and a triangle to make a house.

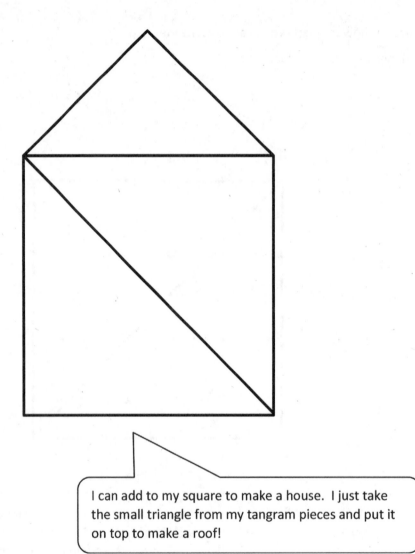

I can add to my square to make a house. I just take the small triangle from my tangram pieces and put it on top to make a roof!

Lesson 5:   Compose a new shape from composite shapes.

Name _____ Date _____

1. Cut out all of the tangram pieces from the separate piece of paper provided.

2. Tell a family member the name of each shape.

3. Follow the directions to make each shape below. Draw or trace to show the parts you used to make the shape.

   a. Use 2 tangram pieces to make 1 triangle.

   b. Use 1 square and 1 triangle to make 1 trapezoid.

   c. Use one more piece to change the trapezoid into a rectangle.

4. Make an animal with all of your pieces. Draw or trace to show the pieces you used.
   Label your drawing with the animal's name.

Lesson 5:   Compose a new shape from composite shapes.

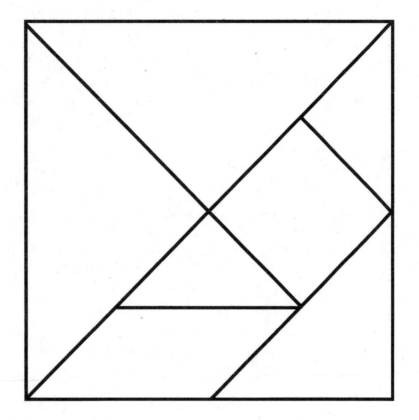

tangram

Use some 3-dimensional shapes to make a structure. Ask someone at home to take a picture of your structure.

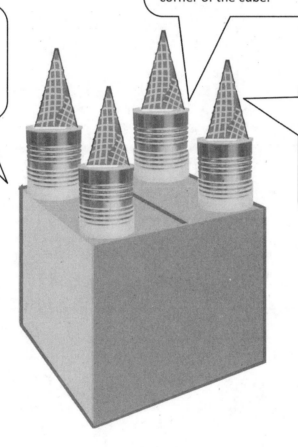

I used 4 cylinders to make the bottom of each tower. I used soup cans for the cylinders. I put each cylinder on a corner of the cube.

I made a castle! I started by putting a big cube on the floor. The cube is a cardboard box!

I used 4 cones to make each tower pointy on the top! I used ice cream cones for the cones. I put each cone on top of each cylinder. Now I have a castle!

Lesson 6:   Create a composite shape from three-dimensional shapes and describe the composite shape using shape names and positions.

Name _____   Date _____

Use some 3-dimensional shapes to make another structure. The chart below gives you some ideas of objects you could find at home. You can use objects from the chart or other objects you may have at home.

Cube	Rectangular prism	Cylinder	Sphere	Cone
Block	Food box: Cereal, macaroni and cheese, spaghetti, cake mix, juice box	Food can: Soup, vegetables, tuna fish, peanut butter	Balls: Tennis ball, rubber band ball, basketball, soccer ball	Ice cream cone
Dice	Tissue box	Toilet paper or paper towel roll	Fruit: Orange, grapefruit, melon, plum, nectarine	Party hat
	Hardcover book	Glue stick	Marbles	Funnel
	DVD or video game box			

Ask someone at home to take a picture of your structure. If you are unable to take a picture, try to sketch your structure or write the directions on how to build your structure on the back of the paper.

Lesson 6:   Create a composite shape from three-dimensional shapes and describe the composite shape using shape names and positions.

149

© 2018 Great Minds®. eureka-math.org

1. Are the shapes divided into equal parts? Write **Y** for yes or **N** for no. If the shape has equal parts, write how many equal parts there are on the line.

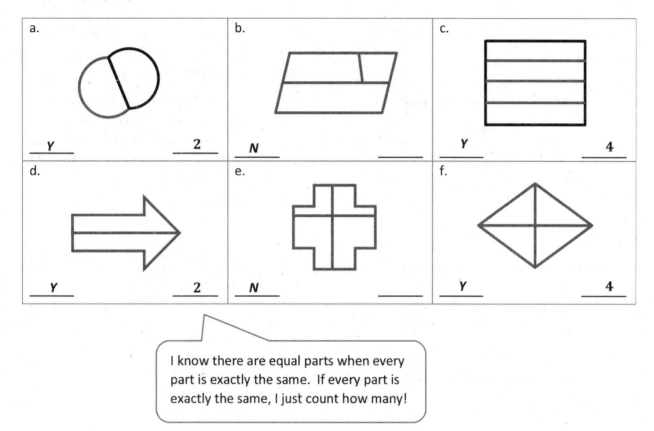

a.

_____Y_____   _____2_____

b.

_____N_____   _____

c.

_____Y_____   _____4_____

d.

_____Y_____   _____2_____

e.

_____N_____   _____

f.

_____Y_____   _____4_____

> I know there are equal parts when every part is exactly the same. If every part is exactly the same, I just count how many!

2. Draw 1 line to make 2 equal parts. What smaller shapes did you make?

> I can make 2 equal parts in different ways. I can make 2 rectangles or 2 triangles.

I made 2 _____*rectangles*_____.

EUREKA MATH

Lesson 7:   Name and count shapes as parts of a whole, recognizing relative sizes of the parts.

© 2018 Great Minds®. eureka-math.org

151

3. Draw 2 lines to make 4 equal parts. What smaller shapes did you make?

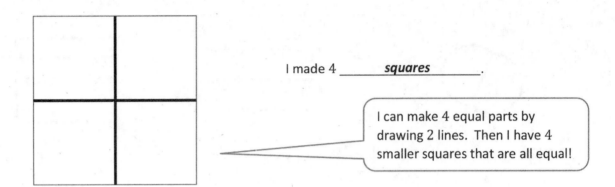

I made 4 _____*squares*_____.

> I can make 4 equal parts by drawing 2 lines. Then I have 4 smaller squares that are all equal!

4. Draw lines to make 6 equal parts. What smaller shapes did you make?

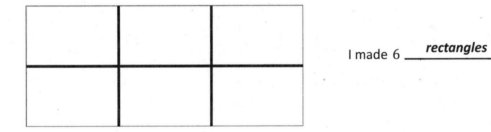

I made 6 ____*rectangles*____.

Lesson 7:   Name and count shapes as parts of a whole, recognizing relative sizes of the parts.

© 2018 Great Minds®. eureka-math.org

EUREKA MATH®

Name _____   Date _____

1. Are the shapes divided into equal parts? Write **Y** for yes or **N** for no. If the shape has equal parts, write how many equal parts there are on the line. The first one has been done for you.

a. **Y** _____ **2** _____	b. _____ _____	c. _____ _____
d. _____ _____	e. _____ _____	f. _____ _____
g. _____ _____	h. _____ _____	i. _____ _____
j. _____ _____	k. _____ _____	l. _____ _____
m. _____ _____	n. _____ _____	o. _____ _____

Lesson 7:   Name and count shapes as parts of a whole, recognizing relative sizes of the parts.

2. Draw 1 line to make 2 equal parts. What smaller shapes did you make?

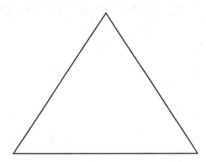

I made 2 _____.

3. Draw 2 lines to make 4 equal parts. What smaller shapes did you make?

I made 4 _____.

4. Draw lines to make 6 equal parts. What smaller shapes did you make?

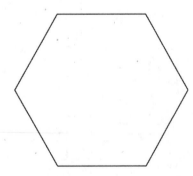

I made 6 _____.

Lesson 7: Name and count shapes as parts of a whole, recognizing relative sizes of the parts.

© 2018 Great Minds®. eureka-math.org

EUREKA
MATH

1. Circle the correct word(s) to tell how each shape is divided.

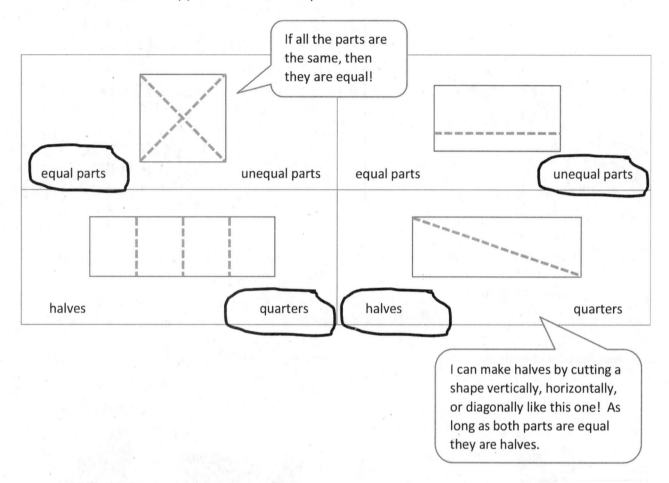

If all the parts are the same, then they are equal!

equal parts   unequal parts   equal parts   **unequal parts**

halves   **quarters**   **halves**   quarters

I can make halves by cutting a shape vertically, horizontally, or diagonally like this one! As long as both parts are equal they are halves.

EUREKA MATH

Lesson 8:   Partition shapes and identify halves and quarters of circles and rectangles.

© 2018 Great Minds®. eureka-math.org

155

2. What part of the shape is shaded? Circle the correct answer.

a.

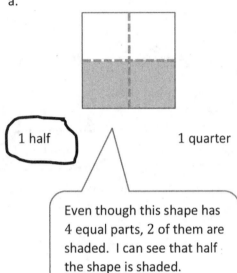

(1 half)   1 quarter

Even though this shape has
4 equal parts, 2 of them are
shaded. I can see that half
the shape is shaded.

b.

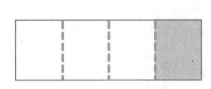

1 half   (1 quarter)

3. Color 1 quarter of each shape.

To color a quarter, I just color
1 of the 4 equal parts!

4. Color 1 half of each shape.

To color a half, I
just color 1 of the
2 equal parts!

To color a half of
this shape I need
to color 2 of the 4
equal parts.

Lesson 8:   Partition shapes and identify halves and quarters of circles and
rectangles.

© 2018 Great Minds®. eureka-math.org

EUREKA
MATH

Name _____ Date _____

1. Circle the correct word(s) to tell how each shape is divided.

a.	b.
equal parts unequal parts	equal parts unequal parts
c.	d.
halves fourths	halves quarters
e.	f.
halves quarters	fourths halves
g.	h.
quarters halves	halves fourths

Lesson 8:   Partition shapes and identify halves and quarters of circles and rectangles.

© 2018 Great Minds®. eureka-math.org

157

2. What part of the shape is shaded? Circle the correct answer.

a.

1 half   1 quarter

b.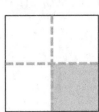

1 half   1 quarter

c.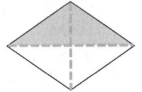

1 half   1 quarter

d.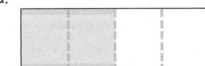

1 half   1 quarter

3. Color 1 quarter of each shape.

4. Color 1 half of each shape.

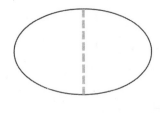   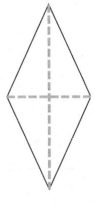

Lesson 8:   Partition shapes and identify halves and quarters of circles and
rectangles.

© 2018 Great Minds®. eureka-math.org

EUREKA
MATH

1. Label the shaded part of each picture as one half of the shape or one quarter of the shape.

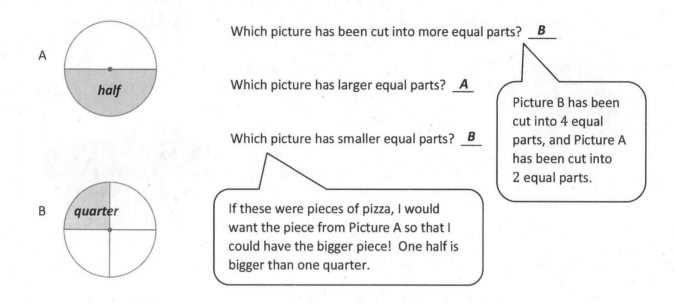

A
half

Which picture has been cut into more equal parts? _**B**_

Which picture has larger equal parts? _**A**_

Which picture has smaller equal parts? _**B**_

Picture B has been cut into 4 equal parts, and Picture A has been cut into 2 equal parts.

B *quarter*

If these were pieces of pizza, I would want the piece from Picture A so that I could have the bigger piece! One half is bigger than one quarter.

2. Write whether the shaded part of each shape is a half or a quarter.

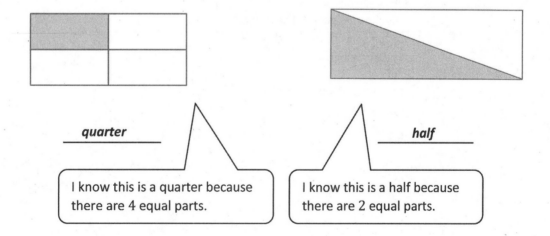

_____quarter_____

I know this is a quarter because there are 4 equal parts.

_____half_____

I know this is a half because there are 2 equal parts.

EUREKA MATH

Lesson 9:   Partition shapes and identify halves and quarters of circles and rectangles.

159

© 2018 Great Minds®. eureka-math.org

3. Color part of the shape to match its label. Circle the phrase that would make the statement true.

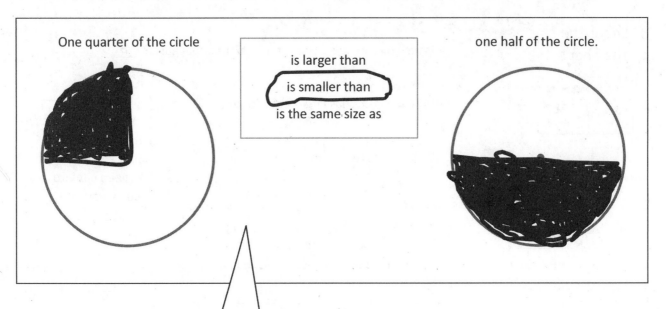

One quarter of the circle

is larger than
~~is smaller than~~
is the same size as

one half of the circle.

A quarter is smaller than a half. If you cut a shape into quarters, you cut it into 4 equal parts. If you cut a shape into halves, you make only 2 equal parts. The more equal parts there are, the smaller the size of the parts.

Lesson 9: Partition shapes and identify halves and quarters of circles and rectangles.

EUREKA
MATH

Name _____   Date _____

1. Label the shaded part of each picture as one half of the shape or one quarter of the shape.

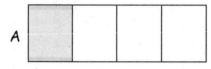

A

Which picture has been cut into more equal parts? ____

Which picture has larger equal parts? ____

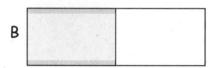

B

Which picture has smaller equal parts? ____

2. Write whether the shaded part of each shape is a half or a quarter.

a. 	b.
_____ _____	_____ _____
c.  	d.  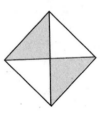
_____ _____	_____ _____

EUREKA MATH

Lesson 9:   Partition shapes and identify halves and quarters of circles and rectangles.

© 2018 Great Minds®. eureka-math.org

161

3. Color part of the shape to match its label. Circle the phrase that would make the statement true.

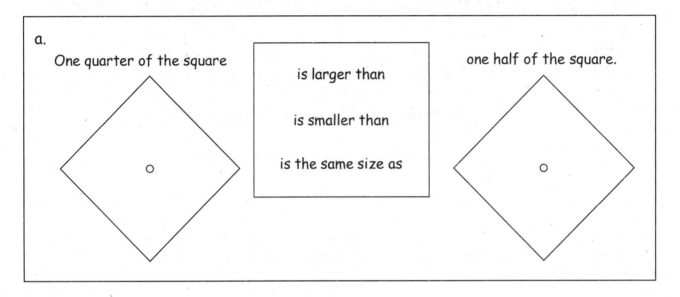

a.

One quarter of the square   is larger than   one half of the square.

is smaller than

is the same size as

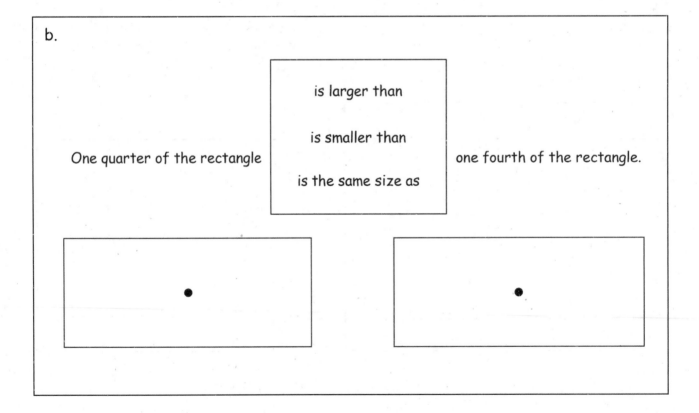

b.

is larger than

is smaller than

is the same size as

One quarter of the rectangle   one fourth of the rectangle.

Lesson 9:   Partition shapes and identify halves and quarters of circles and rectangles.

EUREKA
MATH

1. Match each clock to the time it shows.

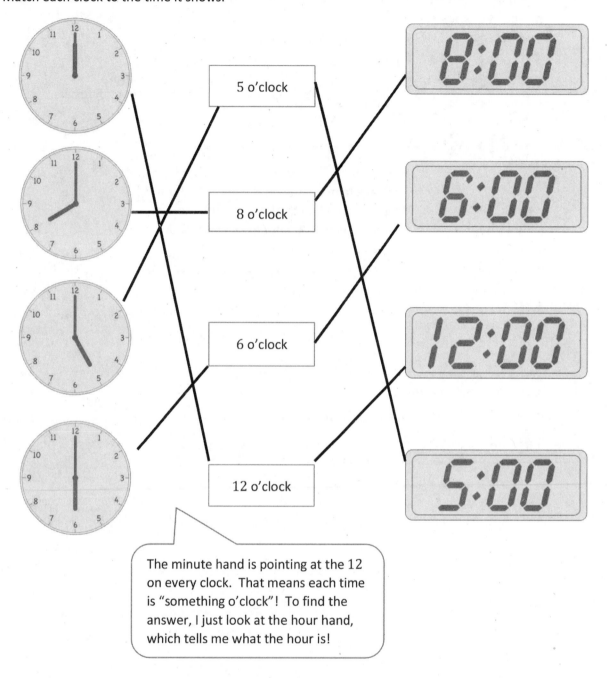

The minute hand is pointing at the 12 on every clock. That means each time is "something o'clock"! To find the answer, I just look at the hour hand, which tells me what the hour is!

Lesson 10:   Construct a paper clock by partitioning a circle and tell time to the hour.

© 2018 Great Minds®. eureka-math.org

163

2. Put the hour hand on the clock so that the clock matches the time. Then, write the time on the line.

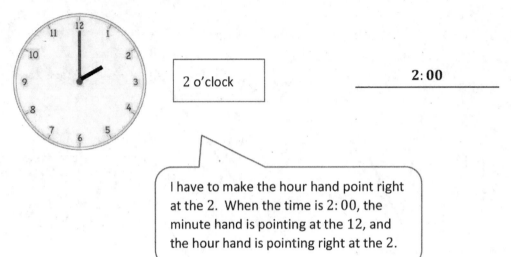

2 o'clock

___2:00___

I have to make the hour hand point right at the 2. When the time is 2:00, the minute hand is pointing at the 12, and the hour hand is pointing right at the 2.

Lesson 10:   Construct a paper clock by partitioning a circle and tell time to the hour.

EUREKA
MATH

Name _____   Date _____

1. Match each clock to the time it shows.

a.

4 o'clock

b.

7 o'clock

c.

11 o'clock

d.

10 o'clock

3 o'clock

e.

2 o'clock

f.

EUREKA
MATH®

Lesson 10: Construct a paper clock by partitioning a circle and tell time to the
hour.

165

© 2018 Great Minds®. eureka-math.org

2. Put the hour hand on the clock so that the clock matches the time. Then, write the time on the line.

a.

6 o'clock

6:00

b.

9 o'clock

_____

c.

12 o'clock

_____

d.

7 o'clock

_____

e.

1 o'clock

_____

Lesson 10: Construct a paper clock by partitioning a circle and tell time to the hour.

EUREKA
MATH

1. Circle the correct clock.

   Half past 12 o'clock

 a.

 b.

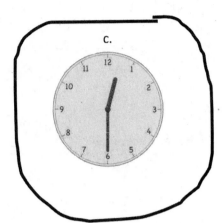 c.

When the time is "half past", the minute hand will always be pointing down, halfway around the clock, at the 6. All these clocks have the minute hand pointing at the 6, so now I just find the clock with the hour hand pointing just past the 12.

The hour hand is not yet at the 1, so I know the hour is still 12.

Lesson 11: Recognize halves within a circular clock face and tell time to the half hour.

167

© 2018 Great Minds®. eureka-math.org

2. Write the time shown on each clock to tell about Henry's Saturday.

Henry wakes up at ____8:30____.

He goes to the park at ____11:30____.

He goes home for lunch at ____1:30____.

He takes a nap at ____2:30____.

> I can check my work by asking myself if my answer makes sense. It wouldn't make sense for Henry to eat lunch at 8:30, for example.

Lesson 11:   Recognize halves within a circular clock face and tell time to the half hour.

EUREKA MATH®

Name _____   Date _____

Circle the correct clock.

1. Half past 2 o'clock

a.   b.   c.

2. Half past 10 o'clock

a.   b.   c.

   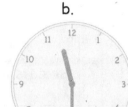   

3. 6 o'clock

a.   b.   c.

4. Half past 8 o'clock

a.   b.   c.

Lesson 11:   Recognize halves within a circular clock face and tell time to the half hour.

© 2018 Great Minds®. eureka-math.org

169

Write the time shown on each clock to tell about Lee's day.

5.  Lee wakes up at _____.	6.  He takes the bus to school at _____.
7.  He has math at _____.	8.  He eats lunch at _____.
9.  He has basketball practice at _____.	10.  He does his homework at _____.
11.  He eats dinner at _____.	12.  He goes to bed at _____.

      Lesson 11:   Recognize halves within a circular clock face and tell time to the half hour.

EUREKA MATH

Write the time shown on the clock, or draw the missing hand(s) on the clock.

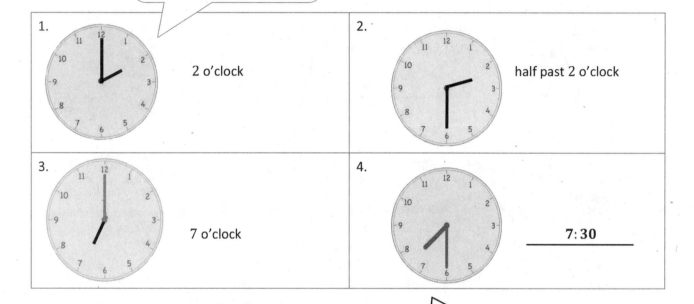

When the time is "o'clock", I draw the minute hand pointing to the 12.

1.   2 o'clock

2.   half past 2 o'clock

3.   7 o'clock

4.   7:30

When the time is "half past" or 30 minutes, I know the minute hand should be pointing halfway around the clock at the 6.

© 2018 Great Minds®. eureka-math.org

5. Match the pictures with the clocks.

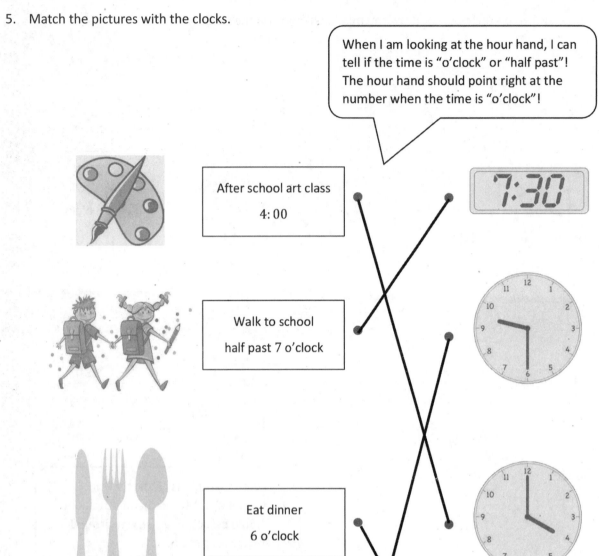

When I am looking at the hour hand, I can tell if the time is "o'clock" or "half past"! The hour hand should point right at the number when the time is "o'clock"!

After school art class
4:00

Walk to school
half past 7 o'clock

Eat dinner
6 o'clock

Math class
9:30

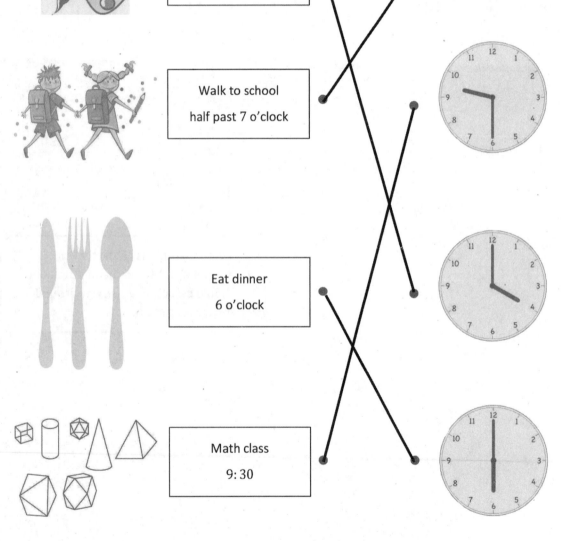

Lesson 12:   Recognize halves within a circular clock face and tell time to the half hour.

EUREKA MATH

Name _____   Date _____

Write the time shown on the clock, or draw the missing hand(s) on the clock.

1. 10 o'clock	2. half past 10 o'clock
3. 8 o'clock	4. _____
5. 3 o'clock	6. half past 3 o'clock
7. _____	8. half past 6 o'clock
9. half past 9 o'clock	10. 4 o'clock

EUREKA MATH®

Lesson 12:   Recognize halves within a circular clock face and tell time to the half hour.

173

© 2018 Great Minds®. eureka-math.org

11. Match the pictures with the clocks.

a.

Soccer practice

3:30

●

b.

Brush teeth

7:30

●

c.

Wash dishes

6:00

●

● **6:30**

d.

Eat dinner

5:30

●

e.

Take bus home

4:30

●

f.

Homework

half past 6 o'clock

●

●

Lesson 12:   Recognize halves within a circular clock face and tell time to the half hour.

EUREKA
MATH

1. Fill in the blanks.

A

B

Clock ___**B**___ shows half past five.

Clock A shows half past 6. This one was easy because it's easy to read the digital clock. It shows "five-thirty."

A

B

Clock ___**A**___ shows seven o'clock.

Both clocks show a time that is "o'clock," but when I look carefully at the hour hands, I see that clock B shows 6 o'clock, and clock A shows 7 o'clock.

Lesson 13:   Recognize halves within a circular clock face and tell time to the half
hour.

175

2. Write the time on the line under the clock.

I also know that if the hour hand is halfway between two numbers, then it will be half past the hour.

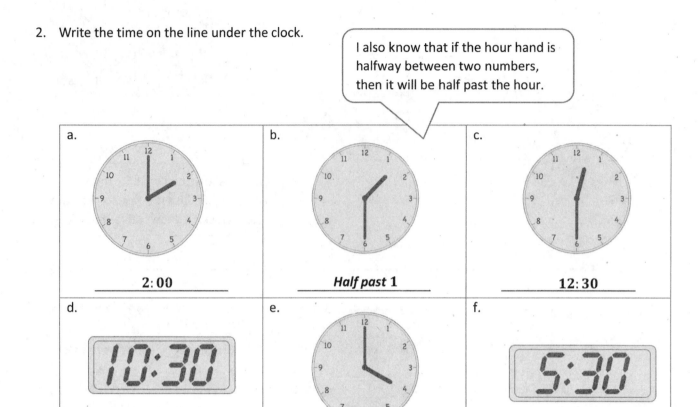

a.

2:00

b.

Half past 1

c.

12:30

d.

Half past 10

e.

4 o'clock

f.

Half past 5

3. Put a check (✓) next to the clock(s) that show 11 o'clock.

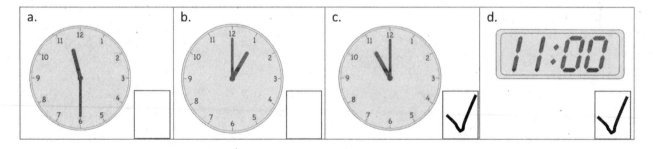

a.

b.

c. ✓

d. ✓

Lesson 13:   Recognize halves within a circular clock face and tell time to the half hour.

EUREKA MATH®

Name _____   Date _____

Fill in the blanks.

1.

A

B

Clock _____ shows half past three.

2.

A

B

Clock _____ shows half past twelve.

3.

A

B

Clock _____ shows eleven o'clock.

4.

A

9:00
B

Clock _____ shows 8:30.

5.

A

B

Clock _____ shows 5:00.

EUREKA
MATH®

Lesson 13:   Recognize halves within a circular clock face and tell time to the half hour.

177

6. Write the time on the line under the clock.

a.	b.	c.
d. 7:30	e.	f.
g.	h. 11:00	i.

7. Put a check (✓) next to the clock(s) that show 4 o'clock.

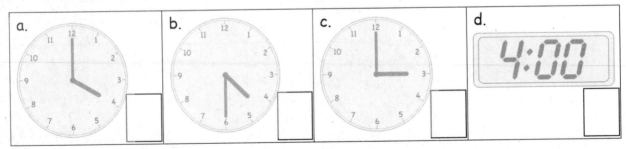

a.   b.   c.   d. 4:00

Lesson 13:   Recognize halves within a circular clock face and tell time to the half hour.

© 2018 Great Minds®. eureka-math.org

EUREKA
MATH®

Grade 1
Module 6

Noah ate 7 jelly beans. His older sister Charlotte ate 15 jelly beans.   How many more jelly beans did Charlotte eat than Noah?

> I can first draw and label a tape diagram to represent the number of jelly beans Noah ate, 7. I can label this tape diagram with the letter *N*.

N | 7 |

C | 7 : ? |

— 15 —

> Next, I can draw and label a second tape diagram right underneath to represent the number of jelly beans Charlotte ate, 15, and label it with the letter *C*. I can see that Charlotte's tape is longer than Noah's because she ate more jelly beans. Drawing and labeling a double tape diagram like this helps me easily compare numbers.

> Noah's tape represents 7, so this much of Charlotte's tape is also 7.

> This part of Charlotte's tape represents how many more jelly beans she ate. I can write a question mark in this part to represent the unknown.

$15 - 7 =$ | 8 |

Charlotte ate 8 *more jelly beans than Noah*.

> Finally, I need to write my statement that matches my story. This will help me check my answer and make sure it makes sense.

> Now I can write a number sentence to find the unknown. There are many strategies to find the unknown. I can count on from 7 to get to 15. I can think of this problem as $7 + ? = 15$ to get 8. But, in this case I choose to use subtraction since it is the most efficient.

Name _____   Date _____

Read the word problem.
Draw a tape diagram or double tape diagram and label.
Write a number sentence and a statement that matches the story.

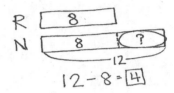

1. Fran donated 11 of her old books to the library.   Darnel donated 8 of his old books to the library. How many more books did Fran donate than Darnel?

2. During recess, 7 students were reading books. There were 17 students playing on the playground. How many fewer students were reading books than playing on the playground?

3. Maria is 18 years old. Her brother Nikil is 12 years old. How much older is Maria than her brother Nikil?

4. It rained 15 days in the month of March. It rained 19 days in April. How many more days did it rain in April than in March?

Lesson 1:   Solve *compare with difference unknown* problem types.

1. Grace used 12 blocks to build a tower. Matt used 4 more blocks than Grace. How many blocks did Matt use?

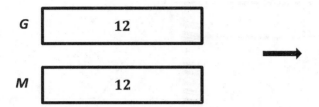

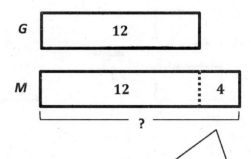

I can draw a double tape diagram to represent the story. First, I can draw a tape diagram that represents the number of blocks, 12, that Grace used to build a tower and label her tape with the letter *G*. Then I can draw a second tape diagram to represent the number of blocks Matt used to build his tower and label it with the letter *M*. Since I don't yet know how many blocks Matt used for his tower, I can begin by drawing and labeling his tape the same size as Grace's.

The story says, "Matt used 4 more blocks than Grace." So, I need to draw an extra part of tape next to Matt's to show that he used 4 more blocks than Grace. The unknown is the total number of blocks Matt used. I can label this with a question mark.

To check that I've drawn and labeled all of the known and unknown information, I can read each part of the story again. As I read, I can touch the part of the double tape diagram that corresponds to what I'm saying.

$12 + 4 = \boxed{16}$

Now I can write a number sentence to help me find the total number of blocks and a statement that answers the question.

Matt used 16 blocks.

2. Susan found 9 fewer seashells than John. John found 13 seashells. How many seashells did Susan find?

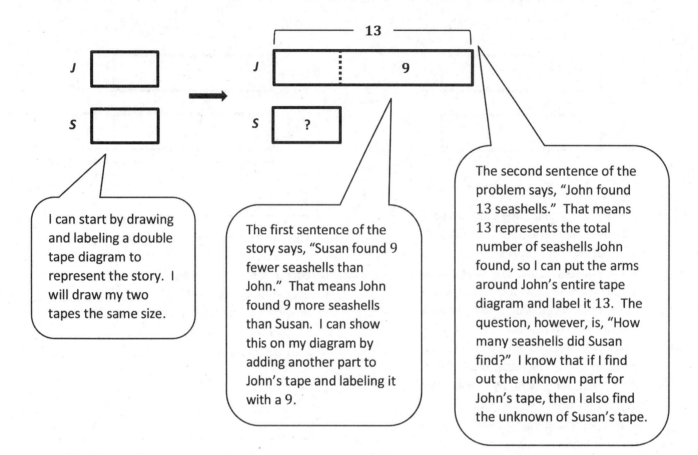

I can start by drawing and labeling a double tape diagram to represent the story. I will draw my two tapes the same size.

The first sentence of the story says, "Susan found 9 fewer seashells than John." That means John found 9 more seashells than Susan. I can show this on my diagram by adding another part to John's tape and labeling it with a 9.

The second sentence of the problem says, "John found 13 seashells." That means 13 represents the total number of seashells John found, so I can put the arms around John's entire tape diagram and label it 13. The question, however, is, "How many seashells did Susan find?" I know that if I find out the unknown part for John's tape, then I also find the unknown of Susan's tape.

$13 - 9 = \boxed{4}$

Susan found 4 seashells.

I can use subtraction to find the missing part. Since John's missing part is 4, Susan's missing part is also 4 because they are the same size. So, Susan found 4 seashells.

EUREKA MATH

Name _____   Date _____

Read the word problem.
Draw a tape diagram or double tape diagram and label.
Write a number sentence and a statement that matches the story.

N [6]
R [6 | 4]
    ?=10
6 + 4 = [10]

1. Kim went to 15 baseball games this summer. Julio went to 10 baseball games.
    How many more games did Kim go to than Julio?

_____

2. Kiana picked 14 strawberries at the farm. Tamra picked 5 fewer strawberries than
    Kiana. How many strawberries did Tamra pick?

_____

3. Willie saw 7 reptiles at the zoo. Emi saw 4 more reptiles at the zoo than Willie.
    How many reptiles did Emi see at the zoo?

4. Peter jumped into the swimming pool 6 times more than Darnel. Darnel jumped in
   9 times. How many times did Peter jump into the swimming pool?

---

5. Rose found 16 seashells on the beach. Lee found 6 fewer seashells than Rose.
   How many seashells did Lee find on the beach?

---

6. Shanika got 12 cards in the mail. Nikil got 5 more cards than Shanika.
   How many cards did Nikil get?

EUREKA
MATH

1. Write the tens and ones. Complete the statement.

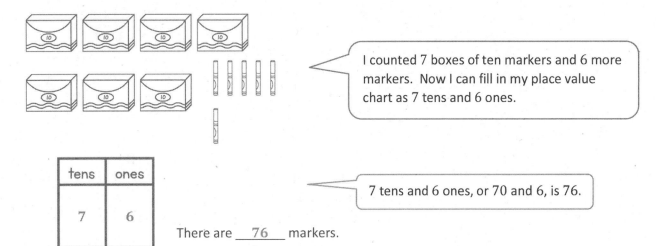

I counted 7 boxes of ten markers and 6 more markers. Now I can fill in my place value chart as 7 tens and 6 ones.

tens	ones
7	6

7 tens and 6 ones, or 70 and 6, is 76.

There are ___76___ markers.

2. Write the number as tens and ones in the place value chart, or use the place value chart to write the number.

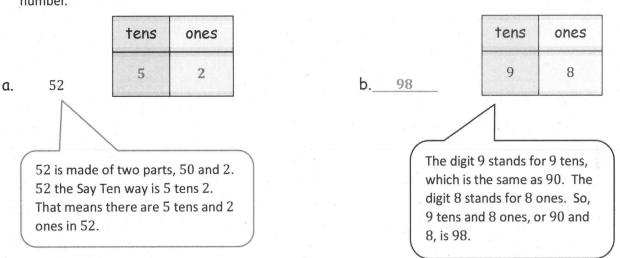

a.   52

tens	ones
5	2

52 is made of two parts, 50 and 2. 52 the Say Ten way is 5 tens 2. That means there are 5 tens and 2 ones in 52.

b.___98___

tens	ones
9	8

The digit 9 stands for 9 tens, which is the same as 90. The digit 8 stands for 8 ones. So, 9 tens and 8 ones, or 90 and 8, is 98.

EUREKA MATH

Lesson 3:   Use the place value chart to record and name tens and ones within a two-digit number up to 100.

189

© 2018 Great Minds®. eureka-math.org

Name _____   Date _____

Write the tens and ones. Complete the statement.

1.

tens	ones

52 = _____ tens _____ ones

2.

tens	ones

_____ = _____ tens _____ ones

3.

tens	ones

There are _____ cubes.

4.

tens	ones

There are _____ cubes.

5.

tens	ones

There are _____ cubes.

6.

tens	ones

There are _____ cubes.

7.

tens	ones

There are _____ carrots.

8.

tens	ones

There are _____ markers.

EUREKA MATH

Lesson 3:   Use the place value chart to record and name tens and ones within a
two-digit number up to 100.

© 2018 Great Minds®. eureka-math.org

191

9. Write the number as tens and ones in the place value chart, or use the place value chart to write the number.

a. 70

tens	ones

b. 76

tens	ones

c. _____

tens	ones
4	9

d. _____

tens	ones
9	4

e. 65

tens	ones

f. 60

tens	ones

g. 90

tens	ones

h. _____

tens	ones
10	0

i. _____

tens	ones
8	3

j. _____

tens	ones
8	0

Lesson 3:   Use the place value chart to record and name tens and ones within a two-digit number up to 100.

EUREKA
MATH™

1. Count the objects, and fill in the number bond and place value chart. Complete the sentences to add the tens and ones.

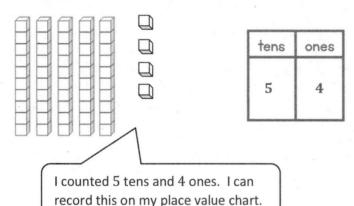

tens	ones
5	4

> I counted 5 tens and 4 ones. I can record this on my place value chart.

> 5 tens and 4 ones is the same as 54. I can break apart 54 as 50 and 4, as shown on my number bond.

> Now I can write addition number sentences that match my number bond. I can either start with the part that represents the tens like I did here or start my number sentence with the ones: $4 + 50 = 54$. I can switch the addends around, and the total is still the same.

$$\underline{\ 50\ } + \underline{\ 4\ } = \underline{\ 54\ }$$

$$\underline{\ 5\ } \text{ tens} + \underline{\ 4\ } \text{ ones} = \underline{\ 54\ }$$

2. Complete the sentences to add the tens and ones.

a. $70 + 4 = \underline{\ 74\ }$

b. 6 tens + $\underline{\ 8\ }$ ones = 68

> I can say this number sentence as "70 more than 4 is 74," or "4 more than 70 is 74," or "70 plus 4 is 74," or "7 tens and 4 ones is 74." These are just some of the many different ways to say this number sentence. This helps me think about numbers flexibly.

EUREKA MATH™   **Lesson 4:**   Write and interpret two-digit numbers to 100 as addition sentences that combine tens and ones.   193

© 2018 Great Minds®. eureka-math.org

Name _____   Date _____

Count the objects, and fill in the number bond or place value chart. Complete the sentences to add the tens and ones.

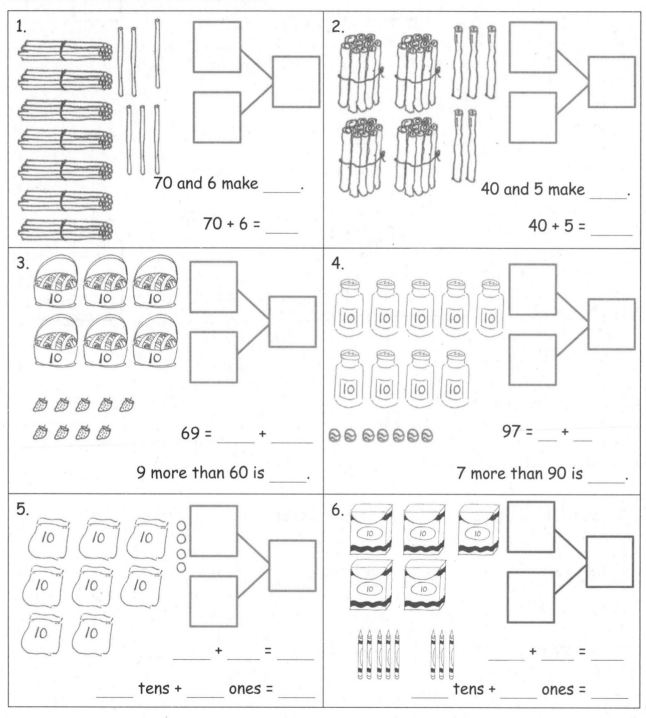

1.

70 and 6 make _____.

70 + 6 = _____

2.

40 and 5 make _____.

40 + 5 = _____

3.

69 = _____ + _____

9 more than 60 is _____.

4.

97 = ___ + ___

7 more than 90 is _____.

5.

_____ + _____ = _____

_____ tens + _____ ones = _____

6.

_____ + _____ = _____

_____ tens + _____ ones = _____

EUREKA MATH™

Lesson 4:   Write and interpret two-digit numbers to 100 as addition sentences that combine tens and ones.

195

© 2018 Great Minds®. eureka-math.org

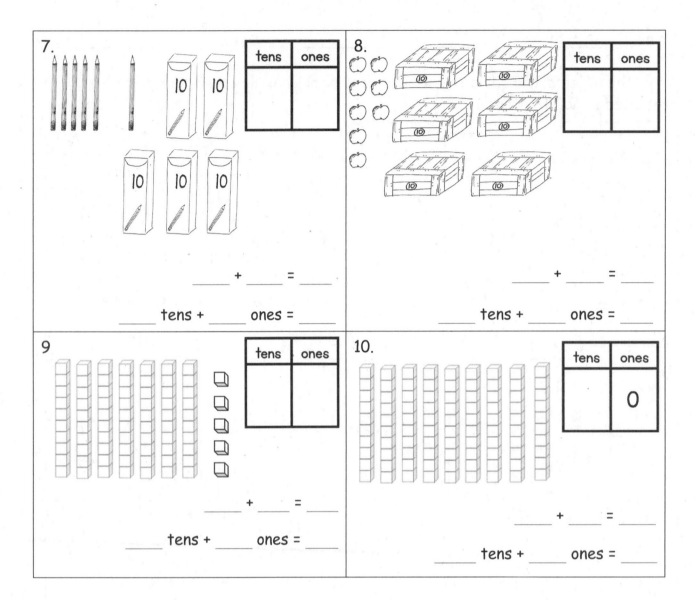

7.

tens	ones

_____ + _____ = _____

_____ tens + _____ ones = _____

8.

tens	ones

_____ + _____ = _____

_____ tens + _____ ones = _____

9

tens	ones

_____ + _____ = _____

_____ tens + _____ ones = _____

10.

tens	ones
	0

_____ + _____ = _____

_____ tens + _____ ones = _____

11. Complete the sentences to add the tens and ones.

a.   80 + 6 = _____

b.   _____ + 7 = 57

c.   9 tens + _____ ones = 95

d.   4 ones + 8 tens = _____

Lesson 4:   Write and interpret two-digit numbers to 100 as addition sentences that combine tens and ones.

© 2018 Great Minds®. eureka-math.org

EUREKA MATH™

1. Find the mystery numbers. Use the arrow way to show how you know.

    a.   1 less than 50 is __49__ .   b.   10 more than 50 is   __60__ .

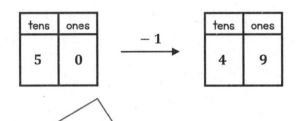

tens	ones
5	0

−1 →

tens	ones
4	9

tens	ones
5	0

+10 →

tens	ones
6	0

There are 5 tens and 0 ones in 50. I can write that in the place value chart on the left. 1 less than 50 is 49. From 50 to 49, I subtracted 1. I can draw an arrow from the first place value chart to the second and write −1 above the arrow. In this case, when I found 1 less, both the tens digit and ones digit changed.

10 more than 50 is 60. From 50 to 60, I added 10. I can draw an arrow from the first place value chart to the second and write +10 above the arrow. Only the tens digit changed this time from 5 tens to 6 tens because we added 10 more. The ones digit did not change.

2. Write the number that is 1 *more*.

    a.   60, __61__

    b.   79, __80__

3. Write the number that is 10 *less*.

    a.   70, __60__

    b.   82, __72__

When I find 1 more or 1 less, sometimes only the ones digit changes, and sometimes both the tens and ones digits change.

I need to read the directions carefully to know when I am adding 1 more, 1 less, 10 more, or 10 less.

Name _____   Date _____

1. Solve. You may draw or cross off (x) to show your work.

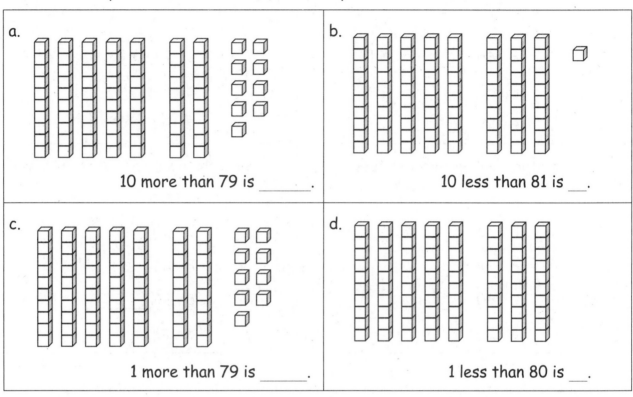

a.

10 more than 79 is _____.

b.

10 less than 81 is ___.

c.

1 more than 79 is _____.

d.

1 less than 80 is ___.

2. Find the mystery numbers. You may make a drawing to help solve, if needed.

a. 10 more than 75 is _____.

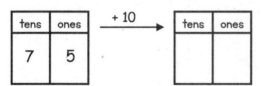

tens	ones
7	5

+ 10 →

tens	ones

b. 1 more than 75 is _____.

tens	ones

→

tens	ones

c. 10 less than 88 is _____.

tens	ones

tens	ones

d. 1 less than 88 is _____.

tens	ones

tens	ones

3. Write the number that is **1 more**.

    a. 40, _____

    b. 50, _____

    c. 65, _____

    d. 69, _____

    e. 99, _____

4. Write the number that is **10 more**.

    a. 60, _____

    b. 70, _____

    c. 77, _____

    d. 89, _____

    e. 90, _____

5. Write the number that is **1 less**.

    a. 53, _____

    b. 73, _____

    c. 71, _____

    d. 80, _____

    e. 100, _____

6. Write the number that is **10 less**.

    a. 50, _____

    b. 60, _____

    c. 84, _____

    d. 91, _____

    e. 100, _____

7. Fill in the missing numbers in each sequence.

    a. 50, 51, 52, _____

    b. 79, 78, 77, _____

    c. 62, 61, _____, 59

    d. 83, _____, 85, 86

    e. 60, 70, 80, _____

    f. 100, 90, 80, _____

    g. 57, 67, _____, 87

    h. 89, 79, _____, 59

    i. _____, 99, 98, 97

    j. _____, 84, _____, 64

Lesson 5:   Identify 10 more, 10 less, 1 more, and 1 less than a two-digit number within 100.

© 2018 Great Minds®. eureka-math.org

EUREKA MATH™

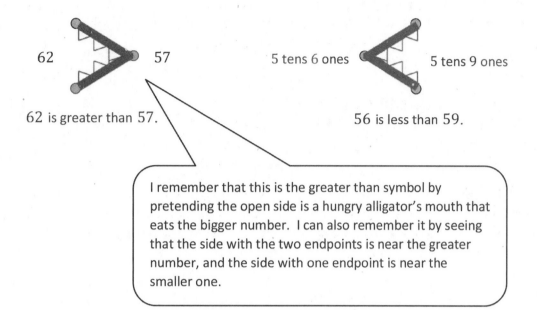

62 > 57

62 is greater than 57.

5 tens 6 ones < 5 tens 9 ones

56 is less than 59.

> I remember that this is the greater than symbol by pretending the open side is a hungry alligator's mouth that eats the bigger number. I can also remember it by seeing that the side with the two endpoints is near the greater number, and the side with one endpoint is near the smaller one.

Circle the correct words to make the sentence true. Use > < or = and numbers to write a true statement.

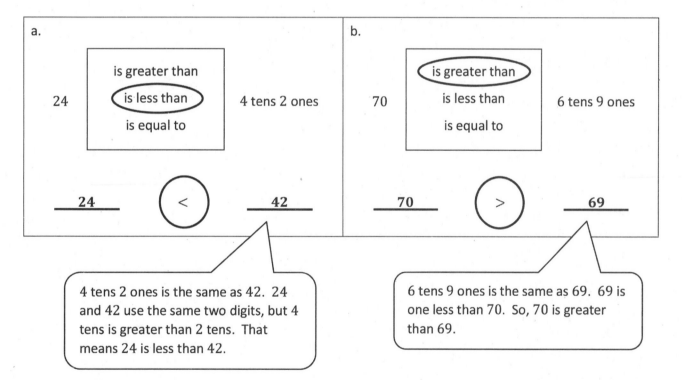

a.

24   is greater than / **is less than** / is equal to   4 tens 2 ones

____24____ (<) ____42____

> 4 tens 2 ones is the same as 42. 24 and 42 use the same two digits, but 4 tens is greater than 2 tens. That means 24 is less than 42.

b.

70   **is greater than** / is less than / is equal to   6 tens 9 ones

____70____ (>) ____69____

> 6 tens 9 ones is the same as 69. 69 is one less than 70. So, 70 is greater than 69.

EUREKA MATH™

Lesson 6:   Use the symbols >, =, and < to compare quantities and numerals to 100.

© 2018 Great Minds®. eureka-math.org

201

Name _____   Date _____

1. Use the symbols to compare the numbers. Fill in the blank with <, >, or = to make
    the statement true.

62   57

62 (>) 57
62 is greater than 57.

5 tens 6 ones   5 tens 9 ones

56 (<) 59
56 is less than 59.

a.	b.
43 ◯ 35	60 ◯ 86

c.	d.
10 tens ◯ 99	5 tens 4 ones ◯ 54

e.	f.
7 tens 9 ones ◯ 9 tens 7 ones	1 ten 3 ones ◯ 31

g.	h.
3 tens 0 ones ◯ 2 tens 10 ones	3 tens 5 ones ◯ 2 tens 17 ones

EUREKA
MATH™

Lesson 6:   Use the symbols >, =, and < to compare quantities and numerals to
             100.

© 2018 Great Minds®. eureka-math.org

203

2. Fill in the correct words from the box to make the sentence true. Use >, <, or = and numbers to write a true statement.

| is greater than | is less than | is equal to |

a.   42 _____ 1 ten 2 ones

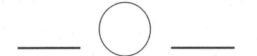

b.   6 tens 7 ones _____ 5 tens 17 ones

c.   37 _____ 73

d.   2 tens 14 ones _____ 4 ones 2 tens

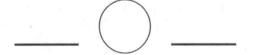

e.   9 ones 5 tens _____ 9 tens 5 ones

Lesson 6:   Use the symbols >, =, and < to compare quantities and numerals to 100.

© 2018 Great Minds®. eureka-math.org

EUREKA
MATH™

1. Complete the chart by filling in the missing numbers.

0	100
1	**101**
2	102
3	103
4	**104**
5	105
6	106
7	**107**
8	**108**
9	109
10	110

I want to be sure to read these numbers without saying *and* between one hundred and the ones place unit. I can read these numbers as, "One hundred one, one hundred two, one hundred three." When I say, "100 *and* 1," it means $100 + 1$, but the name of the number is one hundred one.

2. Compare the 2 columns. What pattern do you notice?

The column on the left counts from 1 *to* 10. *The column on the right counts from* 100 *to* 110. *The pattern is that at* 100 *the numbers start over again from* 0, *only this time you say and write* 100 *first. So, instead of* 1, 2, 3, 4, *it is* 101, 102, 103, 104.

3. Fill in the missing numbers to continue the counting sequence.

a.

97, **96**, 95, **94**

This one is tricky because it is counting down!

b.

99, **100**, **101**, 102

This one is tricky because it is counting to a larger unit. It is going from a 2-digit number to a 3-digit number.

Lesson 7: Count and write numbers to 120. Use Hide Zero cards to relate numbers 0 to 20 to 100 to 120.

205

EUREKA MATH™

© 2018 Great Minds®. eureka-math.org

Name _____ Date _____

1. Fill in the missing numbers in the chart up to 120.

a.	b.	c.	d.	e.
71		91		111
	82		102	
		93		
74				114
	85		105	
		96		116
	87			
			108	
79		99		119
80	90		110	

EUREKA MATH™

Lesson 7: Count and write numbers to 120. Use Hide Zero cards to relate numbers 0 to 20 to 100 to 120.

© 2018 Great Minds®. eureka-math.org

207

2. Write the numbers to continue the counting sequence to 120.

99, _____, 101, _____, _____, _____, _____, _____, _____,

_____, _____, _____, _____, _____, _____, _____,

_____, _____, _____, _____, _____, _____,

3. Circle the sequence that is incorrect. Rewrite it correctly on the line.

a.

| 116, 117, 118, 119, 120 |

b.

| 96, 97, 98, 99, 100, 110 |

4. Fill in the missing numbers in the sequence.

a.

| 113, 114, _____, _____, _____ |

b.

| _____, _____, _____, 120 |

c.

| 102, _____, _____, _____ |

d.

| 88, 89, _____, _____, _____, _____ |

Lesson 7: Count and write numbers to 120. Use Hide Zero cards to relate numbers 0 to 20 to 100 to 120.

© 2018 Great Minds®. eureka-math.org

EUREKA
MATH™

1. Write the number as tens and ones in the place value chart, or use the place value chart to write the number.

tens	ones
7	4

a.   74

74 can be broken apart as 70 and 4, which is the same as 7 tens and 4 ones.

tens	ones
10	9

b.   109

10 tens is the same as 100, and 9 more is 109.

2. Write the number.

a.   10 tens 5 ones is the number _____105_____.

I can read this number as one hundred five, not one hundred *and* five. One hundred *and* five describes $100 + 5$.

b.   11 tens 8 ones is the number _____118_____.

11 tens is the same as 110, and 8 more is 118. I can also show 118 as 10 tens and 18 ones. It is the same number, just written differently.

Lesson 8:   Count to 120 in unit form using only tens and ones. Represent numbers to 120 as tens and ones on the place value chart.

209

© 2018 Great Minds®. eureka-math.org

Name _____   Date _____

1. Write the number as tens and ones in the place value chart, or use the place value chart to write the number.

a. 81

tens	ones

b. 98

tens	ones

c. _____

tens	ones
11	7

d. _____

tens	ones
10	8

e. 104

tens	ones

f. 111

tens	ones

2. Write the number.

a. 9 tens 2 ones is the number _____.	b. 8 tens 4 ones is the number _____.
c. 11 tens 3 ones is the number _____.	d. 10 tens 9 ones is the number _____.
e. 10 tens 1 ones is the number _____.	f. 11 tens 6 ones is the number _____.

EUREKA MATH™

Lesson 8:   Count to 120 in unit form using only tens and ones. Represent numbers to 120 as tens and ones on the place value chart.

211

3. Match.

a.
tens	ones
10	2
●

b.
tens	ones
9	5
●

c.
tens	ones
11	4
●

d.
tens	ones
11	0
●

e.
tens	ones
10	8
●

f.
tens	ones
10	0
●

g.
tens	ones
11	8
●

● | 11 tens 4 ones |

● | 9 tens 5 ones |

● | 11 tens 8 ones |

● | 11 tens 0 ones |

● | 102 |

● | 10 tens 0 ones |

● | 108 |

Lesson 8:   Count to 120 in unit form using only tens and ones. Represent
              numbers to 120 as tens and ones on the place value chart.

EUREKA MATH™

1. Count the objects. Fill in the place value chart, and write the number on the line.

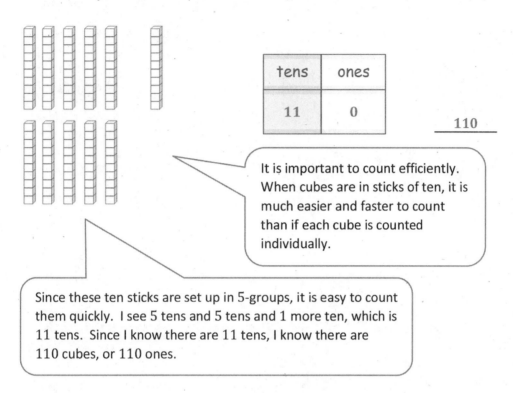

tens	ones
11	0

_____110_____

It is important to count efficiently. When cubes are in sticks of ten, it is much easier and faster to count than if each cube is counted individually.

Since these ten sticks are set up in 5-groups, it is easy to count them quickly. I see 5 tens and 5 tens and 1 more ten, which is 11 tens. Since I know there are 11 tens, I know there are 110 cubes, or 110 ones.

2. Use quick tens and ones to represent the following numbers. Write the number on the line.

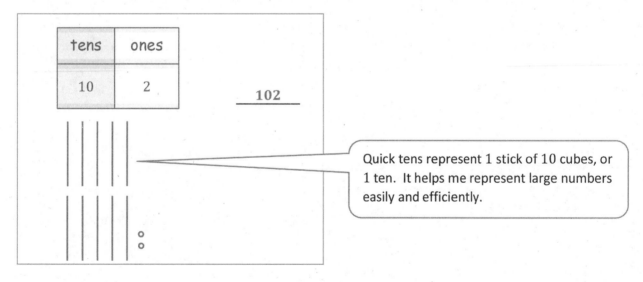

tens	ones
10	2

_____102_____

Quick tens represent 1 stick of 10 cubes, or 1 ten. It helps me represent large numbers easily and efficiently.

EUREKA MATH

Lesson 9: Represent up to 120 objects with a written numeral.

213

© 2018 Great Minds®. eureka-math.org

Name _____   Date _____

Count the objects. Fill in the place value chart, and write the number on the line.

1.

tens	ones

_____

2.

tens	ones

_____

3.

tens	ones

_____

4.

tens	ones

_____

5.

tens	ones

_____

6.

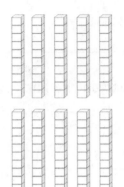

tens	ones

_____

7.

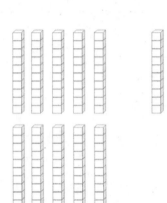

tens	ones

_____

Use quick tens and ones to represent the following numbers.
Write the number on the line.

8. _____

tens	ones
11	0

9. _____

tens	ones
10	5

Lesson 9:   Represent up to 120 objects with a written numeral.

EUREKA
MATH™

1. Complete the number bond or number sentence, and draw a line to the matching picture.

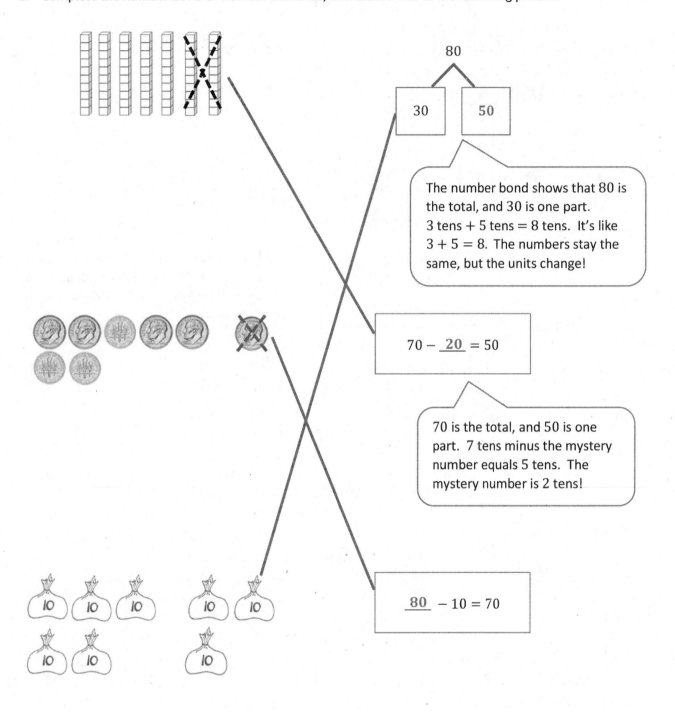

80

30   50

The number bond shows that 80 is the total, and 30 is one part.
3 tens + 5 tens = 8 tens. It's like 3 + 5 = 8. The numbers stay the same, but the units change!

70 − 20 = 50

70 is the total, and 50 is one part. 7 tens minus the mystery number equals 5 tens. The mystery number is 2 tens!

80 − 10 = 70

EUREKA
MATH

Lesson 10: Add and subtract multiples of 10 from multiples of 10 to 100, including dimes.

217

© 2018 Great Minds®. eureka-math.org

2. Count the dimes to add or subtract. Write a number sentence to match the dimes.

$$\underline{ 90 - 30 = 60 }$$

 +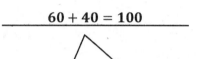

$$\underline{ \mathbf{60 + 40 = 100} }$$

> I can think of $6 + 4 = 10$, to help me. 6 dimes + 4 dimes equals 10 dimes. $60 + 40 = 100$. There is a total of 10 tens!

Lesson 10:   Add and subtract multiples of 10 from multiples of 10 to 100, including dimes.

EUREKA
MATH

Name _____ Date _____

1. Complete the number bond or number sentence, and draw a line to the matching picture.

a.

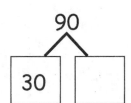

90

30 | ☐

b.

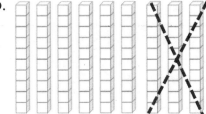

_____ - 40 = 60

c.

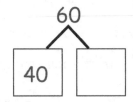

60

40 | ☐

d.

80 - _____ = 60

EUREKA
MATH

Lesson 10: Add and subtract multiples of 10 from multiples of 10 to 100, including dimes.

219

© 2018 Great Minds®. eureka-math.org

2. Count the dimes to add or subtract. Write a number sentence to match the dimes.

a.

    +

_____40 + 20 = _____

b.

_____

c.

+

_____

d.

_____

3. Fill in the missing numbers.

a. 70 + _____ = 90   b. _____ + 30 = 80   c. 100 - _____ = 20

d. 30 + 60 = _____   e. 70 - _____ = 20   f. 20 - _____ = 60

g. _____ - 20 = 60   h. 90 - _____ = 20   h. 50 - _____ = 100

Lesson 10:   Add and subtract multiples of 10 from multiples of 10 to 100, including dimes.

© 2018 Great Minds®. eureka-math.org

EUREKA
MATH

1. Solve using the pictures. Complete the number sentence to match.

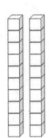 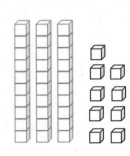

$\underline{20} + \underline{39} = \underline{59}$

I can add 2 tens and 3 tens first. That's 5 tens. I have 9 ones; the ones don't change.

2. Use a number bond to solve.

$40 + 38 = \underline{78}$

30   8

$40 + 30 = 70$

$70 + 8 = 78$

I can break 38 into 30 and 8 with the number bond. I add 40 and 30 first, which is 70, and then add on 8 to make 78.

3. Solve. You may use number bonds to help you.

$23 + \underline{40} = 63$

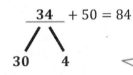 $\underline{34} + 50 = 84$

30   4

I can check my work by drawing a number bond. Since $3 + 5 = 8$, I know that $30 + 50 = 80$. 34 is the missing part because the total, 84, has 4 ones.

I can start at 23 and count on by tens until I get to 63. I count up four tens: 33, 43, 53, 63. 63 is my total!

EUREKA MATH

Lesson 11:   Add a multiple of 10 to any two-digit number within 100.

221

Name _____   Date _____

1. Solve using the pictures. Complete the number sentence to match.

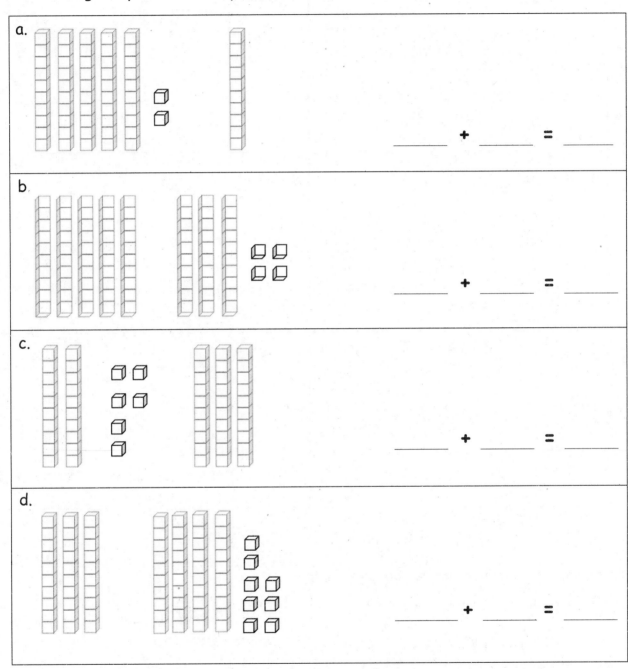

a.

_____ + _____ = _____

b.

_____ + _____ = _____

c.

_____ + _____ = _____

d.

_____ + _____ = _____

EUREKA
MATH®

Lesson 11:   Add a multiple of 10 to any two-digit number within 100.

223

© 2018 Great Minds®. eureka-math.org

$$64 + 30 = 94$$
$$4 \quad 60$$
$$60 + 30 = 90$$
$$90 + 4 = 94$$

2. Use number bonds to solve.

a. 38 + 40 = _____	b. 54 + 30 = _____
c. 46 + 40 = _____	d. 30 + 57 = _____
e. 20 + 68 = _____	f. 25 + 70 = _____

3. Solve. You may use number bonds to help you.

a. 72 + 20 = _____

b. 48 + 50 = _____

c. 46 + _____ = 96

d. _____ + 40 = 87

Lesson 11:   Add a multiple of 10 to any two-digit number within 100.

© 2018 Great Minds®. eureka-math.org

EUREKA MATH

1. Solve.

$38 + 42 = \underline{80}$

    2   40

$38 + 2 = 40$
$40 + 40 = 80$

> I can think about the ones first. Since 38 is close to 40, I can make the next ten! I use a number bond to break apart 42, and then I add $38 + 2$. Then, $40 + 40 = 80$.

2. Solve using number bonds. You may choose to add the ones or tens first. Write the two number sentences to show what you did.

   a. $56 + 43 = \underline{99}$

      40   3

$56 + 40 = 96$
$96 + 3 = 99$

> I can break apart 43 into tens and ones. I can add tens first. So, $56 + 40 = 96$. I can't forget to add the 3 ones: $96 + 3 = 99$.

   b. $25 + 45 = \underline{70}$

  20   5

$45 + 5 = 50$
$50 + 20 = 70$

> This time, I add ones first. When I break apart 25, I see that I can add 5 to 45 to make 50. That's a friendly number! Then I just add 5 tens + 2 tens = 7 tens, or 70.

EUREKA MATH®

Lesson 12:   Add a pair of two-digit numbers when the ones digits have a sum less than or equal to 10.

225

© 2018 Great Minds®. eureka-math.org

Name _____   Date _____

1. Solve.

a. 46 + 22 = _____	b. 74 + 23 = _____
c. 54 + 25 = _____	d. 68 + 31 = _____
e. 45 + 55 = _____	f. 86 + 13 = _____
g. 37 + 52 = _____	h. 47 + 52 = _____

Lesson 12:   Add a pair of two-digit numbers when the ones digits have a sum less than or equal to 10.

227

© 2018 Great Minds®. eureka-math.org

2. Solve using number bonds. You may choose to add the ones or tens first. Write the two number sentences to show what you did.

a. 76 + 23 = _____	b. 45 + 33 = _____
c. 31 + 67 = _____	d. 57 + 32 = _____
e. 58 + 21 = _____	f. 25 + 63 = _____
g. 44 + 55 = ___	h. 47 + 53 = _____

Lesson 12:   Add a pair of two-digit numbers when the ones digits have a sum less than or equal to 10.

© 2018 Great Minds®. eureka-math.org

Solve and show your work.

1. 49 + 24 = __73__

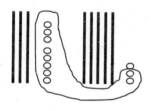
1   23

49 + 1 = 50
50 + 23 = 73

I can think about making the next ten! 49 is close to 50, so I can break apart 24 to add 1 to 49. Then, I add the rest, so 50 + 23 = 73.

2. 38 + 53 = __91__

I can show each number with quick tens and ones. When I look at the ones, I can make another group of ten with 1 leftover. So, I have a total of 9 tens and 1 one, or 91.

3. 25 + 58 = __83__

20   5

58 + 20 = 78
78 + 5 = 83

2   3

I can start with 58 and add 20. To add 78 + 5, I can break apart 5 into 2 and 3. It's easy to solve in my head because 78 + 2 = 80, and 3 more is 83.

4. 67 + 18 = __85__

60   7   10   8

60 + 10 = 70
7 + 8 = 15
70 + 15 = 85

I can break apart both numbers into tens and ones. I add tens first and then ones. I can combine them, so 70 + 15 = 85.

EUREKA MATH®   Lesson 13:   Add a pair of two-digit numbers when the ones digits have a sum greater than 10 using decomposition.   229

© 2018 Great Minds®. eureka-math.org

Name _____   Date _____

1. Solve and show your work.

a. 15 + 26 = _____	b. 46 + 49 = _____	c. 28 + 54 = _____
d. 69 + 13 = _____	e. 69 + 23 = _____	f. 69 + 19 = _____
g. 49 + 43 = _____	h. 57 + 36 = _____	i. 68 + 23 = _____

Lesson 13:   Add a pair of two-digit numbers when the ones digits have a sum
              greater than 10 using decomposition.

231

© 2018 Great Minds®. eureka-math.org

2. Solve and show your work.

a. 34 + 47 = _____	b. 38 + 45 = _____	c. 68 + 23 = _____
d. 39 + 57 = _____	e. 38 + 44 = _____	f. 17 + 76 = _____
g. 68 + 24 = _____	h. 18 + 77 = _____	i. 14 + 67 = _____

Lesson 13:   Add a pair of two-digit numbers when the ones digits have a sum
greater than 10 using decomposition.

© 2018 Great Minds®. eureka-math.org

EUREKA
MATH®

Solve and show your work.

1. $38 + 46 =$ __84__

2   44

$38 + 2 = 40$

$40 + 44 = 84$

> First, I think about making the next ten! I can break apart 46 and add 2 to 38, which makes 40. Then, I add the rest, so $40 + 44 = 84$.

2. $26 + 55 =$ __81__

20   6

$55 + 20 = 75$

$75 + 6 = 81$

5   1

> This time, I can start with 55 and add 20. Then, to add $75 + 6$, I can break apart 6 into 5 and 1 to make a ten. $75 + 5 = 80$, and 1 more is 81.

3. $68 + 17 =$ __85__

60 8   10 7

$60 + 10 = 70$

$8 + 7 = 15$

$70 + 15 = 85$

> I can break both numbers apart into tens and ones. I add tens first and then ones. I can combine them, so $70 + 15 = 85$.

EUREKA
MATH

Lesson 14:   Add a pair of two-digit numbers when the ones digits have a sum greater than 10 using decomposition.

© 2018 Great Minds®. eureka-math.org

233

Name _____ Date _____

1. Solve and show your work.

a. 68 + 21 = _____	b. 59 + 32 = _____
c. 39 + 44 = _____	d. 58 + 36 = _____
e. 76 + 17 = _____	f. 68 + 26 = _____
g. 56 + 39 = _____	h. 58 + 29 = _____

Lesson 14:   Add a pair of two-digit numbers when the ones digits have a sum greater than 10 using decomposition.

© 2018 Great Minds®. eureka-math.org

235

2. Solve and show your work.

a. 39 + 41 = _____	b. 48 + 43 = _____
c. 87 + 13 = _____	d. 59 + 25 = _____
e. 65 + 27 = _____	f. 27 + 67 = _____
g. 49 + 39 = _____	h. 38 + 58 = _____

Lesson 14:   Add a pair of two-digit numbers when the ones digits have a sum greater than 10 using decomposition.

EUREKA MATH

Solve using quick tens and ones drawings. Remember to line up your tens with tens and ones with ones. Write the total below your drawing.

1. $49 + 23 =$ __**72**__

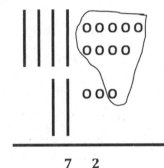

> 49 is 4 tens and 9 ones. 23 is 2 tens and 3 ones. I can line up the tens and the ones to add. I add the ones first. 9 ones and 3 ones is 12 ones. That's 10 and 2. I can circle a new ten and add it to 6 tens. Now I have 7 tens and 2 ones.

      7   2

2. $26 + 68 =$ __**94**__

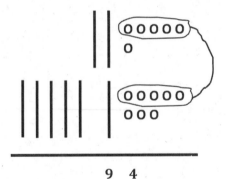

> I make sure to draw each number with quick tens and ones. When I draw the number 68, I put the 6 tens under the 2 tens, and I put the 8 ones under the 6 ones from 26. Look, my 5-group drawings help me to see 10 ones right away!

      9   4

EUREKA MATH®

Lesson 15:   Add a pair of two-digit numbers when the ones digits have a sum greater than 10 with drawing. Record the total below.

© 2018 Great Minds®. eureka-math.org

237

Name _____   Date _____

1. Solve using quick tens and ones drawings. Remember to line up your
    tens with tens and ones with ones. Write the total below your drawing.

a. 39 + 42 = _____	b. 48 + 36 = _____
c. 31 + 48 = _____	d. 47 + 34 = _____
e. 57 + 39 = _____	f. 58 + 27 = _____

Lesson 15: Add a pair of two-digit numbers when the ones digits have a sum
greater than 10 with drawing. Record the total below.

239

© 2018 Great Minds®. eureka-math.org

2. Solve using quick tens and ones. Remember to line up your tens with tens and ones with ones. Write the total below your drawing.

a. 59 + 25 = _____	b. 48 + 42 = _____
c. 39 + 53 = _____	d. 78 + 14 = _____
e. 57 + 25 = _____	f. 69 + 27 = _____

Lesson 15:   Add a pair of two-digit numbers when the ones digits have a sum greater than 10 with drawing. Record the total below.

EUREKA MATH

Solve using quick tens and ones drawings. Remember to line up your drawings and rewrite the number sentence vertically.

1. $49 + 36 =$ __85__

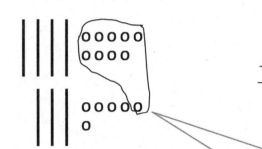

$$\begin{array}{r} 4\ \ 9 \\ +\ \ 3\ \ 6 \\ \hline 1 \\ \hline 8\ \ 5 \end{array}$$

> I can draw 49 as 4 quick tens and 9 ones. So, I write 4 in the tens place and 9 in the ones place. I do the same with 36. I add 4 tens to 3 tens and 9 ones to 6 ones. $9 + 6 = 15$. That's 1 ten 5 ones. Look at where I record the new ten!

8   5

> 9 needs 1 from 6 to get to 10. 10 and 5 is 15.

2. $18 + 78 =$ __96__

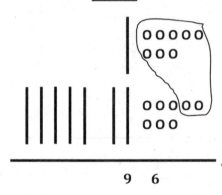

$$\begin{array}{r} 1\ \ 8 \\ +\ \ 7\ \ 8 \\ \hline 1 \\ \hline 9\ \ 6 \end{array}$$

9   6

> When I add 8 ones plus 8 ones, I get 16 ones, which is 1 ten and 6 ones. I record the new ten below the second number in the tens place. 1 ten + 7 tens + 1 ten = 9 tens.

> 8 needs 2 from 8 to get to 10. 10 and 6 is 16.

EUREKA MATH

Lesson 16:   Add a pair of two-digit numbers when the ones digits have a sum greater than 10 with drawing. Record the new ten below.

241

Name _____   Date _____

1. Solve using quick tens and ones drawings. Remember to line up your drawings and rewrite the number sentence vertically.

a. 39 + 45 = _____	b. 64 + 28 = _____
c. 47 + 38 = _____	d. 53 + 27 = _____
e. 38 + 48 = _____	f. 53 + 45 = _____

     Lesson 16:   Add a pair of two-digit numbers when the ones digits have a sum greater than 10 with drawing. Record the new ten below.   **243**

© 2018 Great Minds®. eureka-math.org

2. Solve using quick tens and ones. Remember to line up your drawings and rewrite the number sentence vertically.

a. 79 + 14 = _____	b. 28 + 47 = _____
c. 58 + 33 = _____	d. 19 + 66 = _____
e. 39 + 59 = _____	f. 49 + 48 = _____

Lesson 16:   Add a pair of two-digit numbers when the ones digits have a sum greater than 10 with drawing. Record the new ten below.

© 2018 Great Minds®. eureka-math.org

Solve using quick tens and ones drawings. Remember to line up your drawings and rewrite the number sentence vertically.

1. $58 + 32 =$ __90__

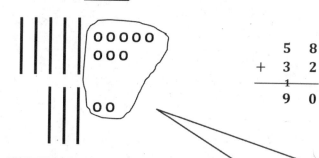

$$\begin{array}{r} 5\ 8 \\ +\ 3\ 2 \\ \hline {\scriptstyle 1} \\ 9\ 0 \end{array}$$

I can draw 58 as 5 quick tens and 8 ones. So, I write 5 in the tens place and 8 in the ones place. I do the same with 32. I add 5 tens to 3 tens and 8 ones to 2 ones: $8 + 2 = 10$. That's 1 ten 0 ones. Look at where I record the new ten!

8 needs 2 to make 10. Now there are 0 ones left.

2. $28 + 49 =$ __77__

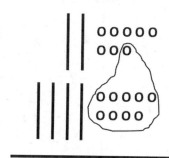

$$\begin{array}{r} 2\ 8 \\ +\ 4\ 9 \\ \hline {\scriptstyle 1} \\ 7\ 7 \end{array}$$

When I add 8 ones plus 9 ones, I get 17 ones, which is 1 ten and 7 ones. I record the new ten below the second number in the tens place. 2 tens + 4 tens + 1 ten = 7 tens.

9 needs 1 from 8 to get to a new 10. Now there are 7 tens and 7 ones.

EUREKA MATH

Lesson 17:   Add a pair of two-digit numbers when the ones digits have a sum greater than 10 with drawing. Record the new ten below.

245

© 2018 Great Minds®. eureka-math.org

Name _____ Date _____

1. Solve using quick tens and ones drawings. Remember to line up your tens and ones and rewrite the number sentence vertically.

a. 49 + 33 = _____	b. 68 + 32 = _____
c. 36 + 43 = _____	d. 27 + 67 = _____
e. 78 + 17 = _____	f. 69 + 28 = _____

Lesson 17:   Add a pair of two-digit numbers when the ones digits have a sum greater than 10 with drawing. Record the new ten below.

© 2018 Great Minds®. eureka-math.org

247

2. Solve using quick tens and ones drawings. Remember to line up your tens and ones and rewrite the number sentence vertically.

a. 29 + 52 = _____	b. 58 + 31 = _____
c. 73 + 26 = _____	d. 67 + 28 = _____
e. 41 + 59 = _____	f. 48 + 45 = _____

Lesson 17:   Add a pair of two-digit numbers when the ones digits have a sum greater than 10 with drawing. Record the new ten below.

EUREKA MATH

Use any method you prefer to solve the problems below.

1. $44 + 23 = \underline{\quad 67 \quad}$

$$
\begin{array}{r}
4\ 4 \\
+\ 2\ 3 \\
\hline
6\ 7
\end{array}
$$

6   7

> I want to draw quick tens and ones to help me solve this problem. The lines represent my tens. The circles represent my ones. I know it is important to carefully line up the tens to tens and the ones to ones.

2. $57 + 23 = \underline{\quad 80 \quad}$

20   3

$$57 \xrightarrow{+20} 77 \xrightarrow{+3} 80$$

> I want to use the arrow way as my strategy. I can break apart 23 into 20 and 3. I can add 20 first and then 3.

3. $48 + 15 = \underline{\quad 63 \quad}$

2   13

$$48 + 2 = 50$$
$$50 + 13 = 63$$

> 48 is so close to 50. I can use the make ten strategy! 48 needs 2 more to make the next ten, 50. I can break apart 15 into 2 and 13. First I can add $48 + 2 = 50$. Then I can add the rest, $50 + 13 = 63$.

Lesson 18:   Add a pair of two-digit numbers with varied sums in the ones, and compare the results of different recording methods.

249

EUREKA
MATH

Name _____   Date _____

Use any method you prefer to solve the problems below.

1. 61 + 15 = _____	2. 16 + 51 = _____
3. 37 + 45 = _____	4. 27 + 46 = _____
5. 58 + 27 = _____	6. 38 + 48 = _____

Lesson 18:   Add a pair of two-digit numbers with varied sums in the ones, and
compare the results of different recording methods.

251

© 2018 Great Minds®. eureka-math.org

Use any strategy you prefer to solve the problems below.

1. $64 + 33 = \underline{97}$

60 4 30 3

$60 + 30 = 90$

$4 + 3 = 7$

$90 + 7 = 97$

I can use double number bonds and break apart BOTH numbers. I can add the tens to the tens,
6 tens + 3 tens = 9 tens, and the ones to the ones,
4 ones + 3 ones = 7 ones. Then, I add all my tens and ones together, 9 tens + 7 ones = 97 ones.

2. $37 + 35 = \underline{72}$

30 5

$37 \xrightarrow{+30} 67 \xrightarrow{+5} 72$

I might want to break apart just one of the numbers. If I break 35 into 30 and 5, I can add 30 first and then add 5. The arrow way is one way I can show my thinking.

3. $38 + 25 = \underline{63}$

$$\begin{array}{r} 3\ 8 \\ +\ 2\ 5 \\ \hline 1 \\ 6\ 3 \end{array}$$

6 3

Another strategy I can use is drawing quick tens and ones.
8 ones + 5 ones = 13 ones. I can bundle 10 of the ones to make 1 ten. I still have 3 ones.
3 tens + 2 tens + 1 ten = 6 tens. There are 6 tens and 3 ones!

Lesson 19: Solve and share strategies for adding two-digit numbers with varied sums.

253

EUREKA MATH

© 2018 Great Minds®. eureka-math.org

Name _____   Date _____

Use the strategy you prefer to solve the problems below.

1. 53 + 22 = _____	2. 23 + 52 = _____
3. 76 + 14 = _____	4. 76 + 16 = _____
5. 55 + 35 = _____	6. 54 + 46 = _____

Use the strategy you prefer to solve the problems below.

7. 49 + 25 = _____	8. 49 + 45 = _____
9. 37 + 37 = _____	10. 37 + 57 = _____
11. 24 + 48 = _____	12. 26 + 68 = _____

Lesson 19:   Solve and share strategies for adding two-digit numbers with varied sums.

EUREKA
MATH

1. Match

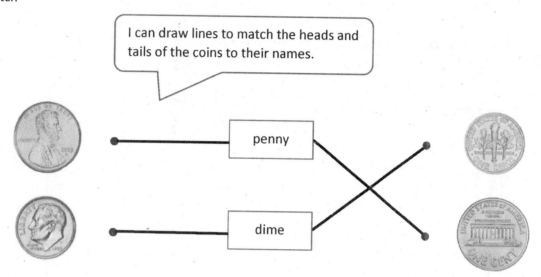

I can draw lines to match the heads and tails of the coins to their names.

penny

dime

2. Cross off some pennies so the remaining pennies show the value of the coin to their left.

A nickel is worth 5 cents. If I cross off 1 penny, the remaining pennies show the value of 1 nickel.

EUREKA MATH

Lesson 20:   Identify pennies, nickels, and dimes by their image, name, or value.
Decompose the values of nickels and dimes using pennies and nickels.

257

© 2018 Great Minds®. eureka-math.org

3.   Marcus has 7 cents in his pocket. Draw coins to show two different ways he could have 7 cents.

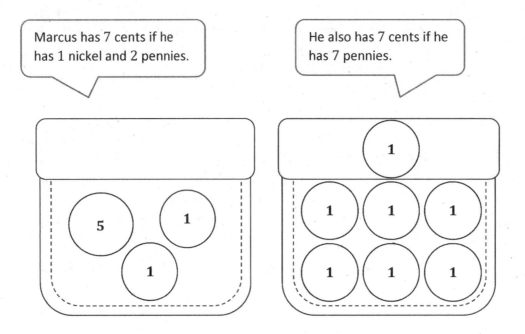

4.   Solve. Draw a line to match the number sentence with the coin or coins that give the answer.

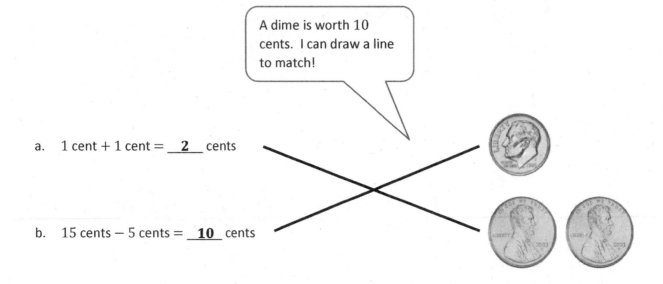

a.   1 cent + 1 cent = __2__ cents

b.   15 cents − 5 cents = __10__ cents

Lesson 20:   Identify pennies, nickels, and dimes by their image, name, or value. Decompose the values of nickels and dimes using pennies and nickels.

© 2018 Great Minds®. eureka-math.org

EUREKA MATH

Name _____   Date _____

1. Match

•   penny   •

•   nickel   •

•   dime   •

2. Cross off some pennies so the remaining pennies show the value of the coin to their left.

a.

b.

EUREKA
MATH®

Lesson 20:   Identify pennies, nickels, and dimes by their image, name, or value.
             Decompose the values of nickels and dimes using pennies and nickels.

259

3. Maria has 5 cents in her pocket. Draw coins to show two different ways she could have 5 cents.

4. Solve. Draw a line to match the number sentence with the coin (or coins) that give the answer.

a. 10 cents + 10 cents = _____ cents   ●   ●

b. 10 cents - 5 cents = _____ cents   ●   ●

c. 20 cents - 10 cents = _____ cents   ●   ●

d. 9 cents - 8 cents = _____ cents   ●   ●

Lesson 20:   Identify pennies, nickels, and dimes by their image, name, or value. Decompose the values of nickels and dimes using pennies and nickels.

© 2018 Great Minds®. eureka-math.org

EUREKA MATH®

1. Use the word bank to label the coins.

| pennies   dimes |

_____ *pennies* _____

I am learning the names and values of coins!

2. Write the value of each coin.

The value of 1 penny is __1__ cent.

3. Your papa said he will give you 1 dime or 1 penny. Which would you take, and why?

I would take 1 dime because it is worth 10 cents. A penny is only worth 1 cent.

I would take the dime because it is more money!

4. Kira has 10 cents in her piggy bank. Which coin or coins could be in her bank? Draw to show two different sets of coins that could be in Kira's piggy bank.

A dime is worth 10 cents. Maybe she has 1 dime.

A nickel is worth 5 cents. She might have 2 nickels.

Lesson 21: Identify quarters by their image, name, or value. Decompose the value of a quarter using pennies, nickels, and dimes.

261

© 2018 Great Minds®. eureka-math.org

Name _____   Date _____

1. Use the word bank to label the coins.

| dimes   nickels   pennies   quarters |

a. _____   b. _____   c. _____   d. _____

2. Write the value of each coin.

a. The value of one dime is _____ cent(s).

b. The value of one penny is _____ cent(s).

c. The value of one nickel is _____ cent(s).

d. The value of one quarter is _____ cent(s).

3. Your mom said she will give you 1 nickel or 1 quarter. Which would you take, and why?

_____

_____

_____

_____

EUREKA MATH®

Lesson 21:   Identify quarters by their image, name, or value. Decompose the value
of a quarter using pennies, nickels, and dimes.

263

4. Lee has 25 cents in his piggy bank. Which coin or coins could be in his bank?

    a. Draw to show the coins that could be in Lee's bank.

    b. Draw a different set of coins that could be in Lee's bank.

Lesson 21:   Identify quarters by their image, name, or value. Decompose the value
of a quarter using pennies, nickels, and dimes.

© 2018 Great Minds®. eureka-math.org

EUREKA
MATH

1. Match the label to the correct coins, and write the value. There may be more than one match for each coin name.

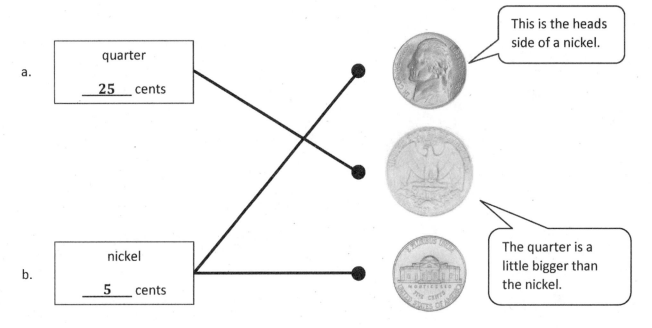

a.
quarter

___25___ cents

This is the heads side of a nickel.

The quarter is a little bigger than the nickel.

b.
nickel

___5___ cents

2. Brian has 4 coins in his pocket, and Larry has 2 coins. Larry has more money than Brian. Draw a picture to show the coins each boy might have.

Hmmm …, Brian has more coins, but Larry has more money. How is this possible?

Brian's Pocket

Larry's Pocket

I have an idea! Maybe Brian has 1 dime and 3 pennies. That's 13 cents. Larry might have 2 dimes, which is 20 cents.
20 is greater than 13, so Larry has more money!

Lesson 22:   Identify varied coins by their image, name, or value. Add one cent to the value of any coin.

265

© 2018 Great Minds®. eureka-math.org

Name _____   Date _____

1. Match the label to the correct coins, and write the value. There will be more than one match for each coin name.

   a.
   ┌─────────────────────┐
   │   **nickel**   │
   │   │
   │ _____ cents   │
   └─────────────────────┘

   b.
   ┌─────────────────────┐
   │   **dime**   │
   │   │
   │ _____ cents   │
   └─────────────────────┘

   c.
   ┌─────────────────────┐
   │   **quarter**   │
   │   │
   │ _____ cents   │
   └─────────────────────┘

   d.
   ┌─────────────────────┐
   │   **penny**   │
   │   │
   │ _____ cent   │
   └─────────────────────┘

     Lesson 22:   Identify varied coins by their image, name, or value. Add one cent to the value of any coin.   267

© 2018 Great Minds®. eureka-math.org

2. Lee has one coin in his pocket, and Pedro has 3 coins. Pedro has more money than Lee. Draw a picture to show the coins each boy might have.

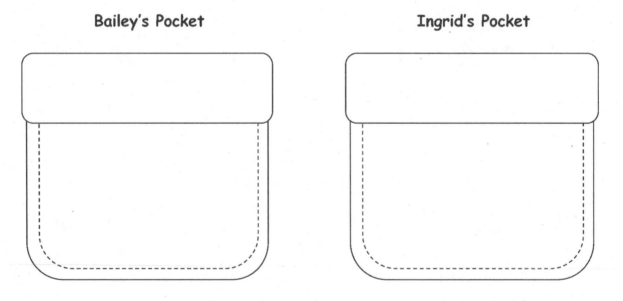

Lee's Pocket   Pedro's Pocket

3. Bailey has 4 coins in her pocket, and Ingrid has 4 coins. Ingrid has more money than Bailey. Draw a picture to show the coins each girl might have.

Bailey's Pocket   Ingrid's Pocket

Lesson 22:   Identify varied coins by their image, name, or value. Add one cent to the value of any coin.

EUREKA MATH

1. Add pennies to show the written amount.

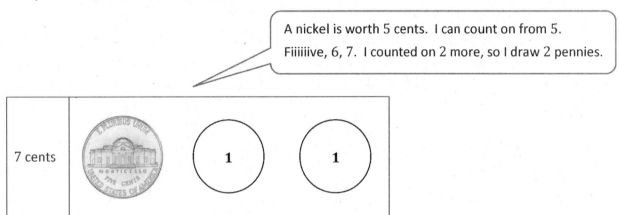

A nickel is worth 5 cents. I can count on from 5.

Fiiiiiive, 6, 7. I counted on 2 more, so I draw 2 pennies.

7 cents

2. Write the value of the group of coins.

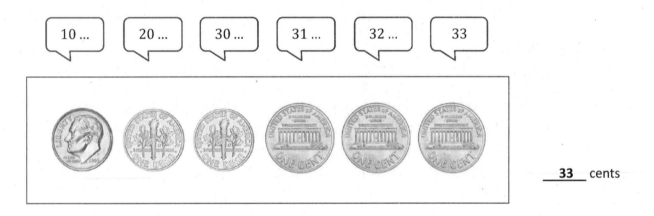

10 …   20 …   30 …   31 …   32 …   33

____33____ cents

EUREKA
MATH

Name _____   Date _____

1. Add pennies to show the written amount.

a. 15 cents	
b. 28 cents	
c. 22 cents	
d. 32 cents	

2. Write the value of each group of coins.

a.

_____ cents

b.

_____ cents

c.

_____ cents

d.

_____ cents

e.

_____ cents

Lesson 23:   Count on using pennies from any single coin.

1. Find the value of each set of coins. Complete the place value chart.

    Write an addition sentence to add the value of the dimes and the value of the pennies.

1 dime = 1 ten.
There are 10 dimes, so there are 10 tens.

1 penny = 1 one.

tens	ones
10	1

$100 + 1 = 101$

10 tens + 1 one is the same as $100 + 1$.
$100 + 1 = 101$

EUREKA MATH

Lesson 24:   Use dimes and pennies as representations of numbers to 120.

273

© 2018 Great Minds®. eureka-math.org

2. Check the set that shows the same amount. Fill in the place value chart to match 100 cents.

There are 8 dimes and 2 pennies, so there are 8 tens and 2 ones: $80 + 2 = 82$.
This set shows 82 cents.

tens	ones
10	0

There are 10 dimes and 0 pennies, so there are 10 tens and 0 ones: $100 + 0 = 100$.
This set shows 100 cents.

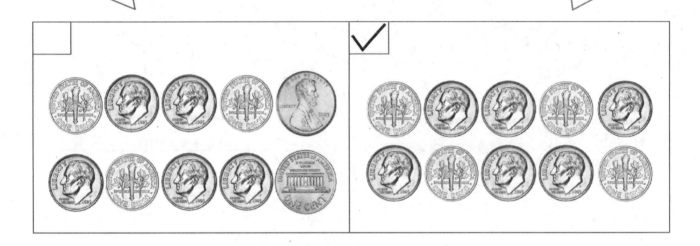

3. Draw 43 cents using dimes and pennies.   Fill in the place value chart to match.

| 10 | 10 | 10 | 10 |

| 1 | 1 | 1 |

tens	ones
4	3

I can make 43 cents with 4 dimes and 3 pennies. That's 4 tens and 3 ones!

Lesson 24:   Use dimes and pennies as representations of numbers to 120.

EUREKA MATH

Name _____   Date _____

1. Find the value of each set of coins. Complete the place value chart.
    Write an addition sentence to add the value of the dimes and the value of the
    pennies.

a.

tens	ones

_____

b.

tens	ones

_____

c.

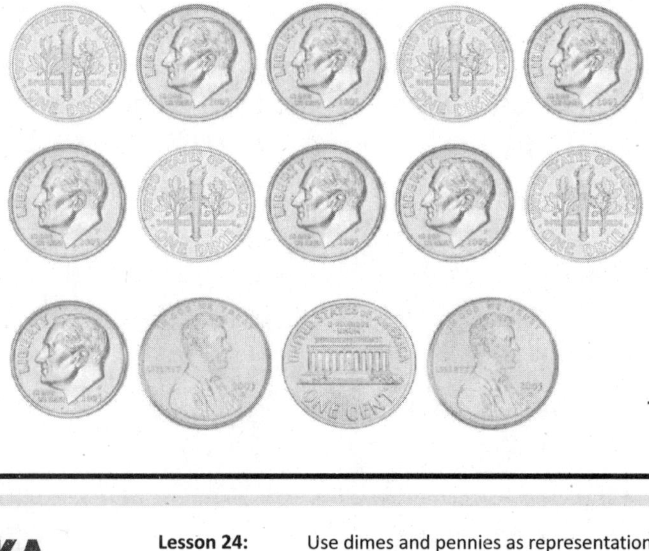

tens	ones

_____

2. Check the set that shows the correct amount. Fill in the place value chart to match.

110 cents

tens	ones

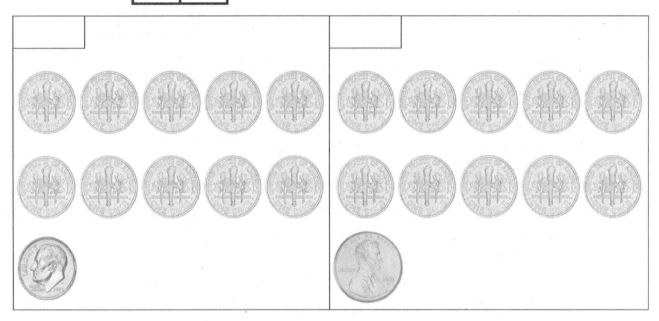

3. a. Draw 79 cents using dimes and pennies. Fill in the place value chart to match.

tens	ones

b. Draw 118 cents using dimes and pennies. Fill in the place value chart to match.

tens	ones

EUREKA MATH

<u>R</u>ead the word problem.
<u>D</u>raw a tape diagram or double tape diagram and label.
<u>W</u>rite a number sentence and a statement that matches the story.

1. Maria used 16 beads to make a bracelet.   Maria used 5 more beads than Kim. How many beads did Kim use to make her bracelet?

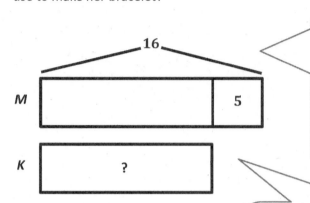

I can draw a double tape diagram to compare Maria's and Kim's beads. I can draw Maria's and Kim's tapes the same length. Since I know they don't have the same amount of beads, I ask myself, Who has more? Maria! She has 5 more beads than Kim. I'll add more to Maria's tape and label it with 5 because she has 5 more beads than Kim.

I can draw arms to include both parts of Maria's tape because the whole is 16. The first part of Maria's tape is equal to Kim's, so if I figure out Maria's first part, I'll know Kim's tape, too!

$16 - 5 = \boxed{11}$

Kim used 11 *beads.*

2. Leo picked 14 strawberries.   Leo picked 4 fewer strawberries than Agnes.   How many strawberries did Agnes pick?

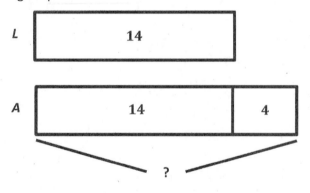

$14 + 4 = \boxed{18}$

Agnes picked 18 strawberries.

I slow down and read every part of the problem carefully. If Leo picked 4 fewer strawberries than Agnes, then Agnes has 4 more than Leo! This is an addition problem, not subtraction!

EUREKA
MATH

Lesson 25:   Solve *compare with bigger or smaller unknown* problem types.

277

© 2018 Great Minds®. eureka-math.org

Name _____   Date _____

Read the word problem.
Draw a tape diagram or double tape diagram and label.
Write a number sentence and a statement that matches the story.

Sample Tape Diagram

N [6]
R [6 | 4]
   ?=10
6 + 4 = 10

1. Julio listened to 7 songs on the radio. Lee listened to 3 more songs than Julio. How many songs did Lee listen to?

2. Shanika caught 14 ladybugs. She caught 4 more ladybugs than Willie. How many ladybugs did Willie catch?

3. Rose packed 3 more boxes than her sister to move to their new house. Her sister packed 11 boxes. How many boxes did Rose pack?

4. Tamra decorated 13 cookies. Tamra decorated 2 fewer cookies than Emi.
    How many cookies did Emi decorate?

5. Rose's brother hit 12 tennis balls. Rose hit 6 fewer tennis balls than her brother.
    How many tennis balls did Rose hit?

6. With his camera, Darnel took 5 more pictures than Kiana. He took 13 pictures.
    How many pictures did Kiana take?

Lesson 25:   Solve *compare with bigger or smaller unknown* problem types.

Read the word problem.
Draw a tape diagram or double tape diagram and label.
Write a number sentence and a statement that matches the story.

1. Ruben has 13 markers. Nashrah has 4 fewer markers than Ruben. How many markers does Nashrah have?

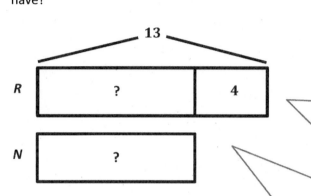

> I can draw a double tape diagram with equal tapes for both Ruben and Nashrah. Since I know they don't have an equal amount of markers, I ask myself, who has more? Since Nashrah has fewer markers, and I know that Ruben has 4 more markers, I'll add more to Ruben's tape and label it with 4 since he has 4 more markers.

$13 - 4 = \boxed{9}$

Nasrah has 9 markers.

> I can draw arms to show Ruben's total, which is 13 markers. The first part of Nashrah's tape is equal to Ruben's, so if I figure out Ruben's first part, I'll know how many markers Nashrah has. I can use subtraction to solve.

2. Emil found 12 leaves on the playground. He found 3 more leaves than Payton. How many leaves did Payton find?

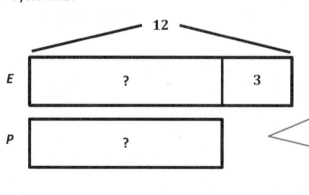

$12 - 3 = \boxed{9}$

Payton found 9 leaves.

> I must read every part of the problem carefully. Sometimes *more* doesn't mean to add! Since Emil found 3 more leaves than Payton, I have to subtract to find out how many leaves Payton found.

EUREKA MATH

Lesson 26:   Solve *compare with bigger or smaller unknown* problem types.

281

© 2018 Great Minds®. eureka-math.org

Name _____   Date _____

Read the word problem.
Draw a tape diagram or double tape diagram and label.
Write a number sentence and a statement that matches the
story.

Sample tape diagram

N ⌈ 6 ⌉
R ⌈ 6 │ 4 ⌉
   ? = 10
  6 + 4 = ⌈10⌉

1. Fatima walks 15 blocks home from school. Ben walks 8 blocks. How much longer is
   Fatima's walk home from school than Ben's?

_____

2. Maria bought a basket with 13 strawberries in it. Darnel bought a basket with 4
   more strawberries than Maria. How many strawberries did Darnel's basket have in
   it?

_____

3. Tamra has 5 books checked out from the library. Kim has 11 books checked out
   from the library. How many fewer books does Tamra have checked out than Kim?

Lesson 26: Solve *compare with bigger or smaller unknown* problem types.

283

© 2018 Great Minds®. eureka-math.org

4. Kiana picked 12 apples from the tree. She picked 6 fewer apples than Willie. How many apples did Willie pick from the tree?

5. During recess, Emi found 16 rocks. She found 5 more rocks than Peter. How many rocks did Peter find?

6. The first grade football team has 12 players. The first grade team has 6 fewer players than the second grade team. How many players are on the second grade team?

Lesson 26: Solve *compare with bigger or smaller unknown* problem types.

EUREKA
MATH

Read the word problem.
Draw a tape diagram or double tape diagram and label.
Write a number sentence and a statement that matches the story.

1. Some children were playing in the gym. 5 children came to join, and now there are 14 children. How many children were in the gym in the beginning?

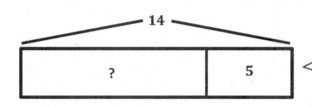

This problem feels tricky because I don't know how many children were playing at first. That's my unknown! It helps when I read one sentence at a time and draw.

$14 - 5 = \boxed{9}$

9 children were in the gym in the beginning.

My drawing shows that I know the whole and one part. I can use subtraction to find out how many children were playing in the beginning. Or, I could have used addition to solve: ___ + 5 = 14.

2. Peter biked for 11 minutes. Belle biked for 7 minutes. How much shorter in time was Belle's bike ride?

P | 11

B | 7 | ?

$7 + \boxed{4} = 11$

Belle's bike ride was 4 minutes shorter.

Since I am comparing this time, I draw a double tape diagram. Since Peter biked for more minutes, his tape is longer than Belle's. I can use addition to solve for the missing part, which is 4 minutes.

Name _____ Date _____

Read the word problem.
Draw a tape diagram or double tape diagram and label.
Write a number sentence and a statement that matches the story.

Sample Tape Diagram

N [6]
R [6 | 4]
?=10
6 + 4 = [10]

1. Eight students lined up to go to art. Some more lined up to go to music. Then, there were 12 students in line. How many students lined up to go to music?

2. Peter rode his bike 5 blocks. Rose rode her bike 13 blocks. How much shorter was Peter's ride?

3. Lee and Anton collected 16 leaves on their walk. Nine of the leaves were Lee's. How many leaves were Anton's?

4. The team counted 11 soccer balls inside the net. They counted 5 fewer *soccer* balls outside of the net. How many soccer balls were outside of the net?

5. Julio saw 14 cars drive by his house. Julio saw 6 more cars than Shanika. How many cars did Shanika see?

6. Some students were eating lunch. Four students joined them. Now, there are 17 students eating lunch. How many students were eating lunch in the beginning?

Lesson 27: Share and critique peer strategies for solving problems of varied types.

1. Teach a family member some of our counting activities. Check all the activities you do together.

   ☐ Happy Count by ones.
   ☒ Happy Count by tens.
   ☒ Count by ones the Say Ten way.
   ☐ Count by tens the Say Ten way.
      First, start at 0, and then start at 7.
   ☒ Movement counting—count while doing squats, arm rolls, jumping jacks, etc.

   > I can practice these fun math games with a family member or friend to keep my math skills sharp over the summer.

2. Write the numbers from 96 to 115.

96	97	98	99	100	101	102	103	104	105

106	107	108	109	110	111	112	113	114	115

3. Count backward by tens from 82 to 2.

   82, __72__, 62, __52__, __42__, __32__, 22, __12__, __2__

   > Practicing a math game like Happy Counting throughout the year has helped me count forward and backward. Look, I can count past 100 by ones and backward by tens! I couldn't do these two things when I started first grade. Now I can do them easily.

EUREKA MATH

Lesson 28: Celebrate progress in fluency with adding and subtracting within 10 (and 20). Organize engaging summer practice.

289

© 2019 Great Minds®. eureka-math.org

Name _____ Date _____

1. Teach a family member some of our counting activities. Check all the activities you do together.

   ☐ Happy Count by ones.
   ☐ Happy Count by tens.
   ☐ Count by ones the Say Ten Way.
   ☐ Count by tens the Say Ten Way. First, start at 0; then, start at 7.
   ☐ Movement counting—count while doing squats, arm rolls, jumping jacks, etc.

2. Write the numbers from 91 to 120:

91		93							

			105						

								119	

1. Count backward by tens from 97 to 7.

   97, _____, 77, _____, _____, _____, _____, _____, _____, _____,

4. On the back of your paper, write as many sums and differences within 20 as you can. Circle the ones that were hard for you at the beginning of the year!

EUREKA MATH

Lesson 28:   Celebrate progress in fluency with adding and subtracting within 10
             (and 20). Organize engaging summer practice.

© 2019 Great Minds®. eureka-math.org

291

Teach a family member your favorite math game during our fluency celebration. Describe what it was like to teach the game. Was it easy? Hard? Why?

I taught my mom how to play the math game Missing Part: Make Ten. I am used to learning how to play the math games from my teacher and then playing with my friends. Teaching my mom was fun, but it was a little bit hard. Even though I know how to play the game, I sometimes forgot to explain some of the important parts to her.

> I can pick a math game from one of our math centers and teach it to one of my family members. I know how to play the game by myself, but sometimes you learn something by teaching it to someone else. It helped me think about making ten when I had to show my mom what we needed to do.

Lesson 29:   Celebrate progress in fluency with adding and subtracting within 10 (and 20). Organize engaging summer practice.

293

© 2019 Great Minds®. eureka-math.org

What did you do in math class today?

Today I decorated a math folder for my math summer packet. I decorated my folder with drawings of all the things I learned in math this year. I drew addition and subtraction number sentences, 5-group drawings, and number bonds. I also drew quick tens, a place value chart, and different two- and three-dimensional shapes. These are just some of the many things I learned in math this year. I will try to practice my summer packet everyday with one of my family members so that I can be ready for math in second grade!

My summer packet includes

- A Lesson 30 Summer Packet.
- Single-sided numeral or 5-group cards.
- 5 Core Fluency Sprints and some other Grade 1 Sprints.
- Core Fluency Differentiated Practice Sets.

Lesson 30:   Create folder covers for work to be taken home illustrating the year's learning.

© 2019 Great Minds®. eureka-math.org

295

Credits

Great Minds® has made every effort to obtain permission for the reprinting of all copyrighted material. If any owner of copyrighted material is not acknowledged herein, please contact Great Minds for proper acknowledgment in all future editions and reprints of this module.